MW00844923

Business as Unusual with SAP®

SAP PRESS is a joint initiative of SAP and Rheinwerk Publishing. The know-how offered by SAP specialists combined with the expertise of Rheinwerk Publishing offers the reader expert books in the field. SAP PRESS features first-hand information and expert advice, and provides useful skills for professional decision-making.

SAP PRESS offers a variety of books on technical and business-related topics for the SAP user. For further information, please visit our website: *www.sap-press.com*.

Thomas Saueressig, Peter Maier

Business as Unusual with SAP®

How Leaders Navigate Industry Megatrends

Editor Emily Nicholls
Copyeditor Yvette Chin
Cover Design Graham Geary
Photo Credit iStockphoto: 1413992340/© filo
Layout Design Vera Brauner, Graham Geary
Production Kelly O'Callaghan
Typesetting SatzPro, Germany
Printed and bound in Germany, on paper from sustainable sources

ISBN 978-1-4932-2389-3

1st edition 2023

Library of Congress Cataloging-in-Publication Control Number: 2022046138

Contents at a Glance

Dear Reader,

This book is a little bit...*unusual* for my colleagues and me.

You may already know SAP PRESS for its far-reaching library of introductions to new software solutions, practical guides offering consultants and business users click-by-click detail, and software programming books with quirky case studies. Books that look beyond the boundaries of the SAP space are a little bit outside our comfort zone.

But from my earliest discussions about this "megatrend" concept with the author team at SAP, I believed that it presented SAP PRESS with an exciting opportunity to reach a new audience. Readers who position their organizations for tomorrow's markets. Leaders who are scanning the horizon and discovering that the industry-altering trends that they once considered at a sedate pace are already here. Innovators who assess, pilot, or commit to new ways of doing business.

For us, the process of publishing this book has already prompted a few departures from our traditional workflows, such as new writing and editing partnerships and a custom interior design to showcase leading voices in the SAP customer and partner space. To borrow a phrase from the book, some of these changes may even become our "next practices."

Bringing *Business as Unusual with SAP* to fruition has been my biggest publishing project of 2022, and could start a new topic area for SAP PRESS. So please reach out and tell me: What did you think about *Business as Unusual with SAP: How Leaders Navigate Industry Megatrends*? Your comments and suggestions are the most useful tools to help us make our books the best they can be. Please feel free to contact me and share your praise or criticism.

Thank you for purchasing a book from SAP PRESS!

Emily Nicholls
Editor, SAP PRESS

emilyn@rheinwerk-publishing.com
www.sap-press.com
Rheinwerk Publishing · Boston, MA

Contents

Appendices 281

Preface

In 2022, we celebrated SAP's 50th anniversary. Fifty years in the software industry means that generations of technology have transformed enterprise computing—from mainframes to client/server architectures, to the internet, and into the cloud. And we see the crests of new technology waves on the horizon.

Our solutions support and enable the daily operations of more than 400,000 customers. And what we consider "business as usual" today once was revolutionary and very un-usual. Real-time computing with SAP R/2 was disruptive when most of the world ran yesterday's data and last month's analytics on homegrown software. SAP R/3, with its scalable client/server computing, displaced the monolithic mainframes. Integrated 24/7 internet-based businesses replaced the usual 9-to-5 weekday schedules. The cloud is paving the way to scalable and adaptable computing power, high-security data management, and continuous innovation, giving companies the agility and resilience they need for an increasingly volatile environment. The pace of change has increased dramatically in the last few years as enterprises navigate one shock after another resulting from COVID-19, the war in Ukraine, climate change concerns, and massive digital transformations.

But some things have not changed over the last 50 years and will remain timeless. We continuously monitor the social, business, and technical megatrends that drive the world. We think about the future of our customers' business in their industries. We co-innovate digital solutions to handle business as unusual for the industry leaders who define intelligent business practices that deliver sustainable business value.

In this book, we examine eight megatrends that are shaping the world of business, the world around us, and the SAP solution portfolio. Analyst and author Vinnie Mirchandani has interviewed many experts and executives of SAP customers and partners and many of our own industry and solution leaders. Vinnie and our industries expert, Tilman Göttke, have compiled a wide range of perspectives to discuss and share a vision of how today's business as unusual will become tomorrow's best practices, enabled by the full range of solutions from SAP and our partners.

Of course, we know that you, dear reader, are observing and evaluating megatrends from your own vantage point, so we look forward to engaging with you as we all explore our shared path into a digital future of intelligent, networked, and sustainable enterprises.

Thomas Saueressig and Peter Maier

Chapter 1
The End of Business as Usual

Generations of business consultants have been making a living by analyzing and re-engineering business processes to find the next level of efficiency that makes their projects (and their fees) worthwhile for clients. If clients are confident that their business model is viable, that their products are competitive, and that their value chain is future-proof, then there is nothing wrong with this focus on how to best run a business within those boundaries.

Meanwhile, strategy consultants take a hard look at the *foundation* of a client's business models to analyze the product portfolio, to investigate market and technology trends, to design the business of the future, and to devise a smooth transformation path to take investors, customers, and employees along for the ride. Nothing is wrong with this approach either; realistically, no book should claim to offer a one-size-fits-all recipe that works for all clients in all circumstances.

In this book, we discuss a range of megatrends that influence many businesses across all industries, shape the strategy of intelligent enterprises, and transform the structure of entire value chains. We hope to start a discussion, not to deliver final answers.

Understanding business and technology megatrends and formulating and testing hypotheses about how they will drive the business is important for SAP's innovation strategy and focus. In times long past, we looked at established business best practices before we developed customizable digital solutions to support and enshrine these business practices. Nowadays, we need to anticipate the best practices of the future—also known as the "next practices"—to build solutions today that will be ready to use tomorrow.

This approach to innovation comes with trade-offs. If we anticipate our customers' future needs correctly, then we'll have the right solutions handy when the demand materializes; otherwise, our solutions will collect dust. But we can easily get caught up in the trap of analyzing markets, trends, and future customer needs so carefully that windows of opportunity for innovative solutions close while we are still in the development phase.

Every innovative company, across all industries, runs the risk of creating solutions that fail because the market has developed differently. Mitigating this risk is important to get a decent return on product investments. However, despite our aspirations, we cannot ensure that 100% of our innovation cases are great customer and commercial successes. In today's hyper-competitive digital market, we need to take calculated risks, so an 80% success rate is already ambitious.

To identify promising innovation areas, we look to our industry and solution experts, but we also talk to customers, partners, and industry analysts to validate our ideas and market insights. In these engagements, we discuss micro-trends that incrementally evolve business practices and related solution requirements without fundamentally changing the playing field or the rules of the game. This level of focus is important to continuously ensure that our solutions are state of the art and meet customer demand and needs. In parallel, we need to take a fundamental look at the global economy, society, technology, and geopolitics to identify and analyze megatrends that influence and transform entire value chains and industry ecosystems.

Any discussion and extrapolation of long-term trends carries the inherent risk that the reality of the future proves today's predictions wrong. Astrologers mitigate this risk by formulating cloudy and unspecific horoscopes and then rely on confirmation bias and on people's tendency to remember the hits but forget the misses. Other authors deal with uncertainty by merely stating facts without taking a step into the unknown by painting the scenarios they expect to evolve from the situations of today and their observable trends.

This book takes a different approach. We discussed a range of megatrends with experts who base today's decisions on their assumptions about fundamental trends and how they will influence business best practices, change business models, and drive the need for new digital solutions that enable future leaders.

These business and thought leaders are experts at taking calculated risks, though they may be proven wrong by a future reality. This expertise also applies to the leaders who have contributed to this book, but the way they talk about megatrends and their conclusions makes all the difference: You'll hear about their reasoning and how they drew their conclusions. This perspective allows you to make up your own mind: which arguments hold water, where your perspective differs, which conclusions you agree with or dispute.

Critical reader engagement begins with selecting the megatrends that form the backbone of this book. We expect each reader to agree with the inclusion of some megatrends, perhaps frown at others, and conclude we missed yet another megatrend or two. If you have this response: Excellent! We've started a discussion that may lead all of us to new insights and into uncharted territories. We have no illusions of giving definitive and holistic answers about the trends driving the future of business, economy, and society; instead, let's see how these trends play out.

1.1 Megatrends and Industries

Traditional books are organized in chapters and sections that are arranged in a sequence. But our selection of megatrends does not suggest that they should be approached and digested in this order. We encourage our readers to follow their preferences and interests, and to pick and choose the megatrends they are curious about and want to engage with:

- **Everything as a Service** is the trend toward developing new business models by moving from making and selling products to delivering product outcomes as a service, thereby transforming revenue streams, product design, and customer engagements.
- **Integrated Mobility** considers the freedom to go places as precious. Mobility that is convenient, safe, and sustainable in urban environments and across continents requires new concepts, business models, and digital technologies.
- **New Customer Pathways** goes beyond a sales transaction and looks at the full customer lifecycle, from making compelling promises to reliably delivering on expectations, thus creating "sticky" customer experiences.
- **Lifelong Health** is an increasingly common concept as smart watches with more sensors than the average intensive-care unit form the visible peak of an emerging health and lifestyle system iceberg centered on the individual patient while integrating a range of industries from healthcare to retail.
- The **Future of Capital and Risk** is important as fintech and insurtech innovators challenge incumbents in the banking and insurance industries by leveraging innovative digital products and services that connect the world of finance with the physical world.
- **Sustainable Energy** is important because finite fossil fuel resources and climate change are good reasons to accelerate the transition to decentralized and renewable energy so we can power a growing economy and global population.
- **Circular Economy** looks at managing resources within our planetary limits. This perspective changes product design and lifecycle management and creates new business opportunities and processes.
- **Resilient Supply Networks** have become important because, for decades, inventory was considered toxic and redundancy, foolish. These days, keeping supply chains lean while making them resilient requires new approaches to combine supply chain transparency with intelligent planning.

These megatrends don't live in isolation but are related to each other. We end up with a web like the one shown in Figure 1.1, which can help you navigate this book by following the threads.

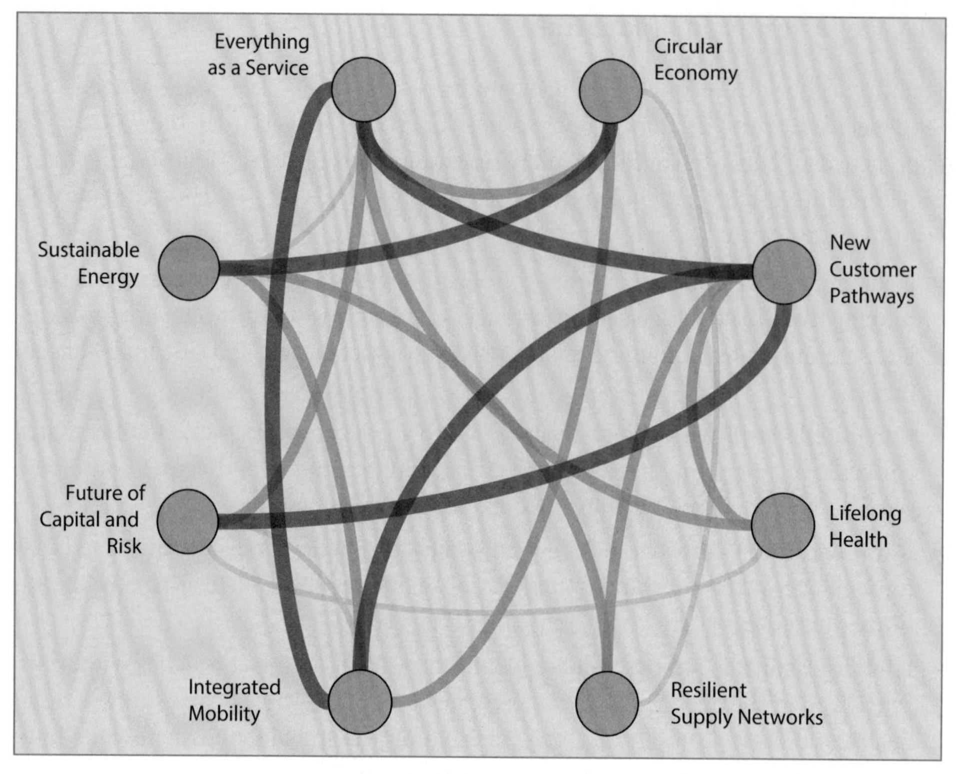

Figure 1.1 Megatrends in a Network with Stronger and Weaker Connections

These megatrends also impact each industry in a specific way. Figure 1.2 shows the relative intensity of each megatrend's impact on many industries, with the darker shading noting a higher impact. You may want to focus on the industries you are most familiar with or have a professional relationship to, but we also encourage lateral reading. Finding out how a megatrend influences a neighboring industry may create new insights and lead to exploring interesting business opportunities.

Let's now take a closer look at each megatrend and their relationships to various industries and to each other. After a short introduction to each, we invite you to immerse yourself in a thought experiment to explore the trend, consider your own exposure, and imagine ways to explore and create business opportunities to capture the potential of megatrends for your organization. If our thought experiments resonate with you, perhaps you can use them with your peers to think outside the box of everyday operations.

Each upcoming section identifies the industries with the biggest exposure to and influence on the megatrend. We also support a non-linear reading experience by closing each section with a short list of closely related megatrends, so you can follow your personal megatrend thread.

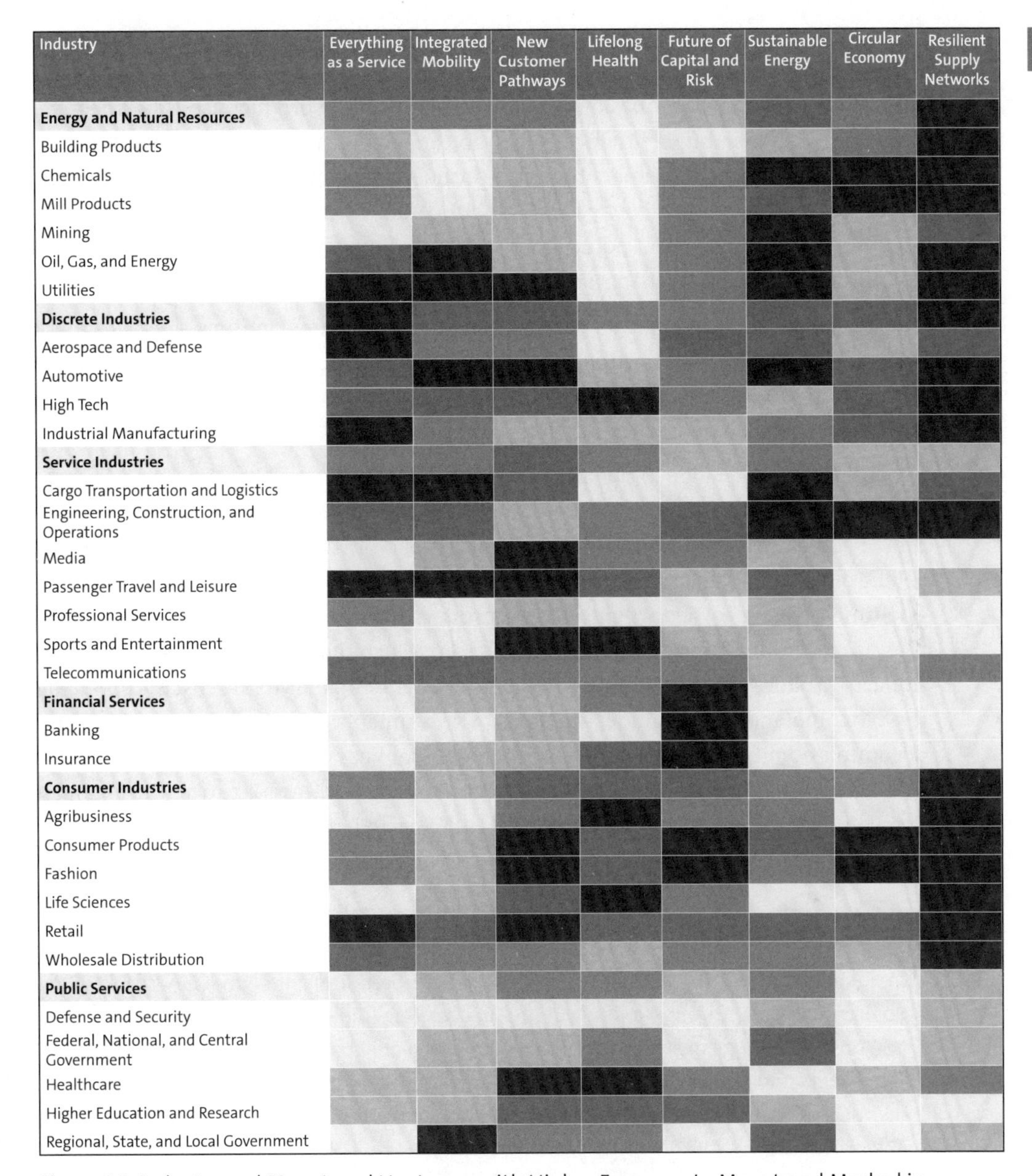

Figure 1.2 Industry and Megatrend Heatmap, with Higher Exposure to Megatrend Marked in Darker Shading

1.2 Everything as a Service

In many industries, making and selling products is the prevailing business model, which defines engineering and design requirements; creates specific cost and revenue streams; and determines product positioning, branding, marketing, sales, and service.

New market entrants find going head to head with incumbents difficult. Because making a cheaper or superior product to win market share is a serious challenge, smart players try to change the rules of game by focusing on the product user's perspective: Why does someone buy a product? What do they expect to get in exchange for their money? Many times, customers buying a product are primarily interested in the outcome that the product delivers: a nice cup of coffee, compressed air, a printed document, safely getting from point A to point B, living in a beautifully appointed apartment—that's what customers ultimately pay for.

Consequently, the next logical step for innovators is to consider how to create these customer outcomes better, with more convenience, predictability, and scalability. The destination of these conversations is often an initiative to deliver and sell a service or an outcome instead of the product that creates the result!

Superficially, this change in perspective looks like a simple and harmless transformation. After all, selling services is by no means a revolutionary idea or business model: Most people don't own trains or planes but instead buy tickets. However, moving to selling outcomes instead of selling products turns the current business upside down: A one-time revenue earned when selling a product transforms into a continuous revenue stream from a subscription for the outcome, such that revenue becomes proportional to product usage. The cost of goods sold (COGS) expands to the cost of producing the outcome and thus includes the cost of consumables, maintenance, service, and decommissioning. The balance sheet becomes asset heavy. Sales teams compare commissions for selling products versus subscriptions. The business risk structure changes, the competitive landscape morphs, and unexpected business opportunities may emerge.

Thought Experiment: Coffee as a Service

Imagine you run a successful business selling your premium coffee machines through your online store, through a lifestyle product retailer, and with your own sales teams. You observe a range of startups trying to convince your customers to trade in their coffee makers and take out a full-service coffee subscription instead. To stay competitive, you decide to offer this option to keep the new competitors at bay and protect your customer base.

The workshop to prepare this move triggers a major transformation program: The coffee machine needs a new connectivity component that counts the cups of coffee made to generate an invoice, trigger the shipment of coffee beans to stay stocked, and schedule service technicians to keep the machine up and running. Your chief financial officer (CFO) runs scenarios of the impact of this model on cash flow, working capital, and revenue and costs to compose a narrative for your investors. Your head of sales needs an enablement package that explains the benefits of subscribing to this coffee model as well as a compensation model to motivate the sales team. Buying your branded coffee machine is no longer the only option for your customers; now, the coffee beans are part of the subscription model, and so, you need new suppliers of high-quality coffee. The chief digital officer already fantasizes about how coffee consumption data can be used beyond ensuring a continuous coffee supply chain. Why not use the data for personalized advertisements from partners or use it to develop a new line of lifestyle products aligned to an individual customer's coffee habits?

Chief information officers (CIOs) leave this kind of workshop with a laundry list of new business capabilities that must be enabled through digital solutions. They need solutions for subscription billing and managing the digital twins of thousands of coffee machines. Third-party service technicians need to be roped into your maintenance network while using artificial intelligence (AI) to predict maintenance needs. And of course, integrated procurement network and supply chain solutions must ensure a continuous flow of coffee beans.

The insight to take away from this thought experiment? What starts as a cool idea on a Post-it note can mushroom into a major transformational program that will receive both support and pushback across your organization—from sales to finance, from engineering to production, from customer service to marketing.

Selling products as a service can make a transformation both strategic and worthwhile. Customer interaction and engagement becomes continuous instead of intermittent. A stream of customer and usage data provides insights for engineering, product management, sales, marketing, and business development. Keeping ownership of production assets also gives full control of the secondary market for used and refurbished equipment. At the same time, flawless service execution is essential to ensure a customer experience that gives them every reason to renew their subscriptions and even more deeply engage with your product, now a service.

Related Megatrends

- **Circular Economy**: Selling outcomes keeps machines and production assets under the ownership of the service provider, which expands the lifespan of equipment and simplifies recycling at end-of-life.
- **Sustainable Energy**: Sustainable, resource-efficient delivery of products as a service is in the best interest of both the service provider and its customers.
- **New Customer Pathways**: Continuously delivering outcomes instead of making one-time equipment sales ensures a deeper level of customer intimacy and thus a deeper understanding of customer needs and preferences.
- **Integrated Mobility**: This megatrend is itself a prime example of transformation, as the automotive industry goes from selling vehicles to selling mobility services.

Key Industries

Industries that traditionally make and sell physical products are great candidates to experiment with new ways to monetize their products as a service, especially the following industries:

- **Aerospace and defense**: This industry has pioneered selling products as a service under a range of labels, most prominently the manufacturers of aircraft engines that sell "power by the hour."
- **Automotive**: Digital trends around mobility enable the delivery of mobility as a service instead of selling or leasing vehicles, finally giving manufacturers direct access to drivers and their passengers.
- **Industrial machinery and components**: Moving to as-a-service models creates customer intimacy, boosts your business's understanding of how its products are used in the field, and unlocks new revenue streams.
- **Chemicals**: Using deep knowledge about their products, especially their safe handling and use, opens up opportunities to monetize the application of these products, for example, by selling coating or painting as a service.
- **High-tech**: Cloud computing is a classic case of delivering products (e.g., hardware and software) as a service, and we see this area expand into the data management and data science space.
- **Professional services**: These intermediaries can easily bridge the gap between the traditional product business and the selling (and buying) of business outcomes instead.
- **Utilities**: Their "products" (i.e., electricity, water, waste management, etc.) are usually viewed as services anyway, but they have ventured into selling renewable energy infrastructure like solar panels and battery storage as subscriptions.

1.3 Integrated Mobility

Going places has been a thing since our ancestors packed up and moved out of Africa. We needed many millennia to arrive in Tasmania, Polynesia, Cape Horn, and Svalbard, but today, we are so accustomed to reaching every city on our planet within a day that unrestrained mobility feels like a human right.

The car parked in front of your house has long been a symbol of freedom and mobility. In the second half of the last century, urban mobility infrastructure has been built for cars moving between suburbia and city centers and for storing cars while owners slept or worked. In a congested metropolis, driving a car has become decidedly lackluster: Owning and driving a car in Manhattan is certainly not fast, convenient, or affordable.

We hope you see where this is headed: Current population density puts us on a collision course with a new mobility challenge. In a few decades, urban centers will host

8 out of 9 people. If we don't reinvent urban mobility, getting around may slow down to a pedestrian pace again.

Thought Experiment: Designing a Mobility System from Scratch

Imagine someone accidentally pushes the mobility system reset button and leaves you in a world without roads, rails, cars, planes, ships, buses, trams, and trucks. How would you build a mobility system for the future?

Maybe you start with the mobility needs of people and businesses. People want to meet up or go to work or learn; they need groceries and want to exercise; or they just enjoy being outside in nature. The COVID-19 pandemic has reinforced that people are social beings, and video conferences and delivery services can't substitute for live human interactions in the long term. Businesses need supplies and workers to make products that must get to customers.

You can probably start designing your mobility and transportation system with the user experience (UX) at the forefront: People want trips from A to B to be safe, convenient, quick, and affordable. On the way, they may want to work, learn, shop, or maybe enjoy some entertainment. The transportation system must move goods around the globe and deliver them on time at the citizen's doorstep or at the business loading dock.

Intuitively, you could try to connect the endpoints of the trip with an integrated mobility solution that begins at the person's door with an individual mobility element (e.g., walking, scooter, bicycle, or car) that seamlessly integrates with group or mass mobility solutions. Scooters, bicycles, cars, and trains are certainly electrical and autonomous; since you have to rebuild the entire infrastructure in our thought experiment, you don't want to mix cars, bicycles, pedestrians, and trucks again. People don't want to own cars that just occupy precious space most of the time, so they just pay for the use or service. Finally, to optimize the system, you'll want all vehicles connected to a network; this could also prevent accidents and reduce traffic congestion.

Connected, autonomous, and shared electric vehicles play a role in the vision of mobility innovators and incumbents, and the idea of mobility as a service has become widely accepted. You'll want to look at a long-term transformation that will involve multiple industries, including energy utilities, car manufacturers, railway operators, airlines, urban transportation systems, and (importantly) the public sector to guide and govern the mobility infrastructure of the future.

Related Megatrends

- **Everything as a Service**: Mobility is a prime example of the duality between owning products (owning shoes, a horse, or a car) and buying services (a ticket on a stagecoach, train, or plane).

- **New Customer Pathways**: People on the move offer plenty of opportunities to engage with them in innovative ways.
- **Circular Economy**: Mobility infrastructure and machines have a massive environmental footprint and offer significant potential to close material cycles, in particular, in the space of e-mobility.
- **Sustainable Energy**: Most mobility today is powered by fossil fuels, and e-mobility only makes sense if powered by renewable energy from batteries and e-fuels.

Key Industries

- **Automotive**: Unsurprisingly, car manufacturers (and their suppliers) are in focus when discussing new mobility paradigms because changing mobility infrastructure, hardware, and services directly impacts billions of people on the planet.
- **Airlines**: This industry plays a key role in long-distance mobility but faces increasing competition over shorter distances from high-speed trains that offer more convenience and lower emissions.
- **Energy and natural resources**: We need materials, fuels, and electricity to power the mobility of the future, and the move to e-mobility creates massive new demands for advanced materials and electrical energy.
- **Travel and transportation**: An integrated mobility system that brings people from their doorsteps to their destinations conveniently and sustainably requires concepts and services that stitch together different modes of travel for new customer experiences.
- **Retail**: The time people spend moving or being moved offers a great range of customer engagement opportunities for retailers, on mobile devices, in their cars, and in mobility hubs like gas stations, train stations, or airports.
- **Public services**: Federal, state, and local government provide a massive part of the mobility infrastructure, services, and regulations, so they have a major influence on the integrated mobility systems of the future.
- **Banking and insurance**: Given the capital requirements and risk management associated with as-a-service business models, there are new opportunities for the finance sector.

1.4 New Customer Pathways

For years, many experts have treated selling to a business and selling to consumers as separate dominions for which different rules apply and different processes and procedures are followed. Today, the insight prevails that *Homo economicus* has been long extinct where our buying, brand, and engagement habits and preferences are concerned. But *Homo economicus* is also on the brink of extinction in the old business world where product specifications, scoring models, sourcing processes, and purchasing procedures try to keep him or her alive. In this chapter, we'll discuss best practices and next practices for effectively designing and running systems that create innovative consumer engagements and pathways.

Thought Experiment: Ruining the Customer Experience

The consumer industries agonize about creating preferences in a world where customer attention is a scarce commodity and innovators continuously transform customer expectations and experiences. We have observed a wide range of best practices and experiments to create the most compelling customer journeys possible, so reminding ourselves of the basics is sometimes difficult.

Let's take a step back and try something different: Challenge yourself to imagine the *worst* possible customer experience—we are confident that you have many anecdotes to share.

Maybe you mentally split the customer experience in two phases: everything that happens before a customer buys a product or service and everything they experience while using the product or service.

On the discovery journey to a product that fits their needs, they can bump into many obstacles: Making sense of electricity or smartphone contracts, insurance policies, or mortgages can be painful. Understanding and comparing the functions and features of a kitchen appliance or a power drill can take them down rabbit holes. You may give up on this leg of the journey and turn to an expert who will consult them on picking the product that best fits their requirements—or that is on the consultant's bonus plan. So even before they swipe their credit card or sign a contract, you can easily imagine how to create an evil customer experience.

Your next set of opportunities to create painful customer experiences comes during fulfillment: delivering the wrong product in the wrong size in the wrong quantities to the wrong location, or making the customer wait at home for a technician that can't find the house or read a calendar.

Returning a product, ordering maintenance services, or processing an insurance or warranty claim all offer a wide range of opportunities to make a customer's life miserable.

Three things can be learned from this thought experiment:

- A proven way to destroy the customer experience is making a promise and then consistently breaking it.
- Destroying the customer experience is a lifetime challenge; it's not over when the deal is closed.
- There is no discernible difference between consumers and the people who represent business customers.

People have expectations and experience the customer situation regardless of their roles as consumers or business representatives. Creating compelling customer experiences focusing on customer lifetime value and based on a seamless transition from making compelling promises to fulfilling expectations requires a strong bridge between the "front office" and the fulfillment and service functions in the "back office." If at this point you frown and wonder why we put the front and back offices in quotes, you may already be challenging the idea of making this distinction at all: All functions related to making and keeping customer promises are part of shaping a holistic customer experience.

Related Megatrends

- **Integrated Mobility**: Moving people efficiently frees up time to spend productively or recreationally, offering new customer engagement opportunities for a range of industries.
- **Future of Capital and Risk**: Even traditional banks and insurers have a fully virtual portfolio of financial services that can be digitally attached in innovative ways to the phases of their customers' lives, from kindergarten to nursing home.
- **Everything as a Service**: This model disrupts and transforms traditional revenue patterns, cash flows, and financial risks, thus creating the need for innovative financial products to enable and support this transformation.

Key Industries

The first industry that comes to mind in this context is of course the retail industry, which is the traditional transition point between the world of business and the world of consumers. Reviewing the industries that SAP serves, you'll find that many have—or seek to have—direct contact, interaction, and engagement with you and me. The following industries may be less obvious:

- **Higher education and research**: This industry engages with a population of students that will form the economic backbone of society. In a world of rapidly changing skill requirements, the concept of earning a college degree and being done with learning is outdated. Keeping students around for a lifetime of learning (not just as alumni sending their own kids to school) opens new opportunities for higher education institutions.
- **Utilities**: Although not a traditional consumer industry, this industry has great potential to engage with customers around compelling and innovative energy and mobility products and services.
- **Healthcare**: This industry is adjusting its patient focus to the individual and their desire to live long and healthy lives, so prevention and health consulting over a range of digital and in-person channels holds big promises—for people and the health ecosystem.

1.5 Lifelong Health

For most of human history, health and healthcare have revolved around rather analog and unscientific interactions between doctors and their patients. After all, our inner workings were poorly understood, and the world of microbes was unknown until Antonie van Leeuwenhoek discovered them in the 17th century. Since the days of pioneers like Edward Jenner, Louis Pasteur, Robert Koch, and Alexander Fleming, life sciences and healthcare have made unbelievable progress and and became part of the scientific enterprise. In 1953, James Watson and Francis Crick published the structure of DNA, which encodes all life on this planet, the starting point for a deeper understanding of how life and evolution work on a molecular level. Today, DNA analysis is used in healthcare for personalized medicine and treatment or to assess the risks of contracting specific genetically influenced diseases. We've even gone beyond the analysis: Scientists routinely cut, copy, and paste strands of DNA to modify the genetic code. Emmanuelle Charpentier and Jennifer Doudna won the 2020 Nobel Prize in Chemistry for developing the CRISPR-Cas9 genome-editing technology.[1] mRNA technology programs cells to make proteins, most famously the SARS-CoV-2 spike protein, so that our immune system gets ready to defend us against the actual virus.

Living a long and healthy life is a top priority for us humans, so unsurprisingly, the world spends 10% of its gross domestic product (GDP) on health, with the United States the leader among high-income countries at 17%.[2] The full range of digital technologies can be applied to facilitate long and healthy lives, including diagnostics; research and development for new medicine and medical devices; and the manufacturing supply chain and distribution for medical products. The field of diagnosing and treating patients is central to all healthcare, but the use of digital technologies is mushrooming to lifestyle areas including the "quantified self" movement, which has created a vast market for digital devices that collect and compile an increasing set of health- and habit-related data.

Thought Experiment: Exploiting Your Digital Twin

Imagine you continuously collect health- and activity-related data with your smart watch, intelligent scale, bicycle ergometer, sensor-studded running shoes, and workout machines: You'll end up with a massive dataset about your blood oxygenation, heart rate, fitness metrics, sleep and activity patterns, locations, mobility, blood sugar, and much more. Throw in your electronic patient record, and ultimately, you'll have created your own digital twin from all this health and activity data.[3]

Now, imagine you're participating in a design thinking workshop at an innovative startup that wants to create a business from providing goods and services based on the analysis of your digital twin. After a few minutes, you see Post-it notes listing things like life and health insurance, nutrition recommendations and recipes with food delivery options, training schedules, sports equipment, fashion, mattresses, food supplements, gym subscriptions, and much more.

Of course, smart startups have read business 101 on "how to create a platform business" and will try to convince you to share your digital twin, so they can give you health insights for a small subscription fee (or even for free) while they reserve the right to share your data with their business or even government partners. The value proposition is strong: They promise you sound advice about how to live a healthier and longer life, and you get all the goods and services you need to turn advice into action.

Would you sign up?

Ten years ago, this thought experiment would have been science fiction. Today's reality uses these concepts and technologies to create digital twins of machines, buildings, and people. The idea of viewing man as a machine is centuries old—just think of Leonardo da Vinci's *Vitruvian Man*, Mary Shelley's *Frankenstein*, or Fritz Kahn's *Man Machine*! Now this concept is going to the next digital level.

At the same time, the life sciences and healthcare industries are using digital technologies to lay the foundation for a holistic system that fulfills the promise of lifelong health. The following examples illustrate the range of initiatives and how broad the transformation of the complex health system could be:

- Completely electronic health and patient records are a reality in some countries like Estonia, while they are many years in the future in most other countries.
- Clinical trials for new medicines are complex and expensive and could use efficient digital solutions to ensure timely supply of medicine to the participants, collect trial results, and minimize the number of dropouts.
- Personalized medicine integrates patient-specific data in the manufacturing and supply chain. The system needs to ensure that the syringe goes into the right vial for the right patient, and that the patient's data is treated according to regulations.
- Collaboration along the pharmaceutical supply chain must be fast and seamless if we want to effectively fight pandemics. When and where the next pandemic will hit is hard to predict, so supply chain structures can't be static but instead need to work as a "popup" system of laboratories, research and development facilities, manufacturing facilities, and distribution networks. BioNTech's mobile BioNTainer production system for mRNA-based vaccines is a frontrunner in this space.[4]
- Drug counterfeiting is a massive problem in many parts of the world, harming people and damaging the brand and business of pharmaceutical companies. Enabling all partners along the supply chain to verify that medicine is genuine can be achieved through digitally registered packages.

SAP is working with the life sciences and healthcare ecosystem to address these complex priorities that shape toward providing better and more economically viable healthcare for more people in an aging population.

Related Megatrends

- **New Customer Pathways**: The health and lifestyle domain offers a wide range of product and service opportunities to shape new customer engagements and experiences.
- **Everything as a Service**: Maintaining your health is a lifelong challenge that is perfectly compatible with health solutions from innovative providers who seek continuous and evolving customer engagements based on their products delivered as services.

Key Industries

The healthcare and life sciences industries are obvious focus points for a wider ecosystem of industries that promise to help people live long and healthy lives, such as the following industries:

- **Healthcare**: Lifestyle-related chronic diseases have become the primary cause of death in a growing number of countries, so a focus on prevention protects people and contains healthcare costs for both individuals and society.
- **Life sciences**: Personalized medicine and treatment, in combination with advanced diagnostics and continuous health monitoring, can change research and development, manufacturing, and supply chains for pharmaceuticals and medical devices.
- **Consumer products**: The "quantified self" with personalized nutrition and a digitally controlled lifestyle enables new product categories and new direct-to-consumer engagement opportunities.
- **Retail**: With their frequent consumer interactions and increasingly detailed digital customer profiles, retailers can segment their customer bases to approach them with tailored offerings to promote and reward healthy lifestyles while connecting them with the corresponding product portfolios.
- **Insurance**: Health and life insurance companies are changing into consultants nudging customers to adopt healthier lifestyles, resulting in closer customer relationships and better combined ratios.

1.6 Future of Capital and Risk

In the movie *Cabaret*, Liza Minelli's character, Sally Bowles, found that money makes the world go 'round. Fifty years later, this principle hasn't changed, but much else has changed in the way capital is raised and allocated, risks are transferred, and payments are handled in our global economy. Every agreement around goods or services generates a parallel financial transaction that models risks, cash flows, payments, and settlements. These transactions can be as simple as paying cash for a latte at the coffeeshop or as complex as executing a multi-billion-dollar acquisition.

Traditionally, banks and insurers played the key role in operating the world's financial system, governed by national and international regulation. A quick Google image search for "bank building" finds 19th-century marble buildings and 20th-century skyscrapers that epitomize power, trust, and stability. Of course, the financial crisis of 2008–2009 has shown just how misplaced confidence in our financial system can be. Maximizing profit while minimizing risk incentivizes players to bend the rules and slip through loopholes that regulators try to fix, most often after the damage has been done.

The 21st century has seen a new species of players in the financial arena that have carved out chunks of the incumbents' business and challenged the foundation of our financial system with cryptocurrencies that live outside the control of central banks. These fintechs operate at the edge or even outside of regulation, arguing that regulation may be effective to protect the incumbents' outdated business models but has proven ineffective at curbing the misbehavior of some players or preventing systemic failures. Regardless of where we stand in this discussion, fintechs have undisputedly stirred up the financial and capital market. This market is so dynamic because financial products and services can be 100% digital—from design to sales to operations to fulfillment. Thus, the speed of innovation and time to market is only limited by the product designer's creativity and the time required to define and implement new products and services in the banking and insurance systems.

Thought Experiment: No-Touch Financial Services?

Since financial products and services are digital in nature, can you imagine designing and delivering these products and services without any human intervention?

Let's start with the easier part: If you're like most people, you only speak to a bank employee or insurance agent when a (hopefully rare) exception occurs. You probably (and rightfully) assume that most of the day-to-day processing of financial services is already performed by machines. Now, expand this view to exceptions, like failing payment transactions or issues with insurance claims. Rules govern how to deal with these issues, rules that can be encoded into a system.

Moving on to potentially less rule-based activities, like selling a financial product, test yourself: When was the last time you visited the branch of a bank or met with an insurance agent to buy a financial product? Only bank robbers enter the branch of a bank they are not affiliated with. A chatbot powered by AI could engage effectively with you on a website or in a virtual world environment like the metaverse. Buying simple property and casualty insurance is already just a little checkbox away when you book a trip or shop online. Even more complex products like mortgages, securities, or wealth management services are formalized and probably follow regulatory rules and sales commissions schemes.

How about product design? Can you imagine being a digital product manager? Let's take another leaf out of nature's playbook: Evolution works by random mutation and non-random selection. Our financial services products might consist of a set of properties that can vary, including the target customer segment and sales and service channels. Let's start with an initial product and set it free. You could then have properties change randomly over time and let customers perform some non-random selection. Our products are fully digital, so we can have a mutation occur at every touchpoint with a customer. Our "survival of the best-fit" algorithm selects the successful mutations in terms of profitability, adoption, cross-selling, or any other performance metric we want to use.

Not too long ago, creating a new financial product took established bank and insurers many moons and even more software developers to tweak decades-old, homegrown software written in arcane programming languages. Creative fintechs can give the incumbents a run for their money (pun intended), but the big dogs won't just fold and leave the world of capital and financial services to the new kids on the block. Incumbents can create their own fintechs and take them off leash, allowing them to disrupt their own business models and to experiment with innovative ways to interact and engage with new customer segments.

Related Megatrends

- **Lifelong Health**: Countries with aging populations spend double-digit percentages of their gross domestic product (GDP) on healthcare, and people's desire to live a long and healthy life, in financial security, offers huge potential for innovative banking and insurance services.
- **Integrated Mobility**: New forms of movement unlock potential for a wide range of banking and insurance services.
- **New Customer Pathways**: Understanding the lifelong journey of people and enterprises, in terms of finance and risk, is critical to strategically maximize how much of the "wallet" is captured by banking and insurance services.
- **Everything as a Service**: This trend fundamentally transforms cash flows and risk allocations for both the providers and consumers of products that are monetized as services.

Key Industries

- **Banking**: Every business transaction creates corresponding financial transactions, and the digitization of business models requires equally digital and innovative financial services. Fully digital fintechs can give the incumbents a run for their money.

- **Insurance**: The business of quantifying, allocating, and monetizing risk is transforming, driven by digital technologies and technology-influenced risk categories, for example, in the spaces of mobility and health.
- **Manufacturing**: To deliver products as a service (e.g., industrial manufacturing, chemical, aerospace and defense manufacturing, automotive), companies are integrating financial services in their service offerings and actively using innovative ways to manage capital and risk.

1.7 Sustainable Energy

When we talk about energy, "sustainable" has become a synonym for "emission-free" or "net-zero." If we take a single human lifespan as a yardstick, even our youngest children won't themselves see the world *run out* of fossil fuels in their lifetimes. But if we take a step back and think in terms of centuries or even millennia, we must increasingly accept that depleting fossil hydrocarbon resources in a matter of decades or in a handful of centuries will make our great-grandchildren mad at us—and rightfully so.

So, "sustainable energy" must be interpreted as "energy that can be generated for a very long time (i.e., many millennia) within planetary boundaries (i.e., clean atmosphere without excessive levels of greenhouse gases and no toxic waste on land or in the oceans)." For example, consider that planting trees binds atmospheric carbon dioxide in the short term, but the trees decompose after a few decades and release carbon dioxide back into the atmosphere. This means that offsetting fossil fuel emissions by planting a tree doesn't make sense. Nuclear fission is emission free but uses a finite resource, uranium, while nuclear fusion has been 30 years in the future for many decades.

Our global economy runs on energy, which is not going to change. Meanwhile, fossil fuels are finite and their consumption drives climate change. Consequently, the development of sustainable energy has been pioneered in recent decades and has dramatically accelerated, driven by commitments to fight climate change by curbing greenhouse gas emissions. Many enterprises have committed to becoming "carbon neutral." For example, SAP has promised to be carbon neutral in its own operations in 2023 and carbon neutral along its full supply chain by 2030.[5] We are not alone with this initiative; you would be hard pressed to find any corporate website that lacks some kind of climate- or energy-related initiative.

The mission of making our energy system carbon neutral is closely linked to fighting climate change with its catastrophic consequences for humankind and nature—historic droughts, deadly floods, rising sea levels, mass extinctions. In the meantime, governments and businesses have committed to take climate action by avoiding, reducing, and compensating for their greenhouse gas emissions.

Reinventing the global energy system is a massive undertaking that involves all industries, all nations, and all people on this planet. Today, this statement may seem trivial, but we are witnessing a paradigm change in the classic structure of the energy system, which had been based on large-scale, centralized generation and downstream distribution for usage by commercial and residential customers.

Thought Experiment: Designing a Sustainable Energy System

Imagining a sustainable energy system that operates within planetary boundaries and that is not exposed to geopolitical forces is a big task—even if just for a thought experiment.

Let's begin by defining key design criteria that would facilitate sustainability:

- **Net-zero emissions**
  Solar and wind energy are obvious, but nuclear energy or burning fossil fuels with carbon capture and storage (CCS) or compensation for the emitted greenhouse gases check this box.
- **Long-term viability**
  Burning fossil fuels (and arguably nuclear fission) depletes the planet's resources. Solar, wind, and water energy will be available until our sun burns out. Fuels made from organic matter and the ever-evasive nuclear fusion are also sustainable because biomass is circular and the hydrogen isotopes going into a fusion reaction can be produced by a range of physical processes.
- **Immunity to geopolitics**
  Geopolitical environments change over years, centuries, and millennia—and sometimes days. But if we want to start building a sustainable energy system, we must look at the political world map to find vast spaces where wind and solar energy can be harvested at scale with a low risk of being held hostage in the burgeoning energy conflicts between countries and political systems.
- **Transportation and storage**
  Energy must be transported from where and when it was generated to where and when it is needed, so we must always think about covering distance and time. Electricity is difficult to transport over very long distances and (currently) expensive to store for a long time.

If we allow ourselves a timeline of a few decades, we can start to think about the energy system of the future.

Imagine yourself as a consultant to the president of a fictitious country with friendly neighbors and huge deserts not too far from the equator. Maybe your roadmap to become a new key player in the energy economy will have the following simple elements:

- **Capital**
  Find a few hundred billion dollars in countries that are desperate for clean, affordable, and reliable energy.

- **Project**
  Build massive solar and wind farms in your deserts. Integrate sustainable farming in the shade of solar panels.
- **Power-to-X**
  Make the electricity produced transport friendly and easy to store by converting solar and wind energy into chemicals, including hydrogen (H_2), ammonia (NH_3), hydrocarbons like acetylene (C_2H_2), methanol (CH_3OH), and other fuels and feedstocks for the chemicals industry.

Most countries either have vast deserts or are in politically stable regions; only a few benefit from both. While we work on building a planetary political and economic system that acknowledges that we all share one small planet, everybody needs to take a decentralized approach towards a sustainable energy system.

A few energy initiatives for individuals and communities include the following:

- Decentrally collecting solar energy on residential and commercial roofs to charge cars and power homes can generate electricity without claiming more land. Many large energy players and startups already play in this vibrant space by offering products and services to unlock this potential at scale. Installing solar panels on the roof is something almost any homeowner can do.
- Wind turbines must be big to be effective, and big means that they are very visible and very expensive. So wind energy requires community or city-level consensus and financing.
- Saving energy is a big factor as well: Energy efficient residential and commercial buildings can go a long way in saving natural gas and electricity. How about setting your air conditioning and heating so that you wear t-shirts in the summer and wool sweaters in the winter, not the other way around?
- Offshore windfarms are beyond the scope of many communities but are good for energy yield and effective against the "not in my back yard" (NIMBY) attitude.

We don't have any more excuses—the transition to renewable energy works one electron and one molecule at a time.

Related Megatrends

- **Integrated Mobility**: Our mobility and transportation systems currently run mostly on fossil energy, which doesn't even include the energy demand for building and maintaining mobility infrastructures and hardware.
- **Resilient Supply Networks**: Supply chains need reliable, affordable, and sustainable energy to make and move goods.
- **Circular Economy**: A circular and sustainable global economy must keep materials in closed reuse loops, powered by equally sustainable energy.

Key Industries

All industries that provide or use energy on a large scale are directly exposed to the market and to the regulatory forces of transitioning to a sustainable energy system, especially the following industries:

- **Oil and gas**: Still providing more than half of the world's primary energy, this industry's key competence in developing and operating large-scale capital projects will be instrumental in creating pathways to a sustainable energy system.
- **Mining**: Contributing more than a quarter of the world's primary energy in the form of coal, this industry also digs up and processes the minerals we need to build the global sustainable energy system.
- **Automotive, travel, and transportation**: Moving people and goods around make up a significant share of the world's energy demand: 28% of final energy is used by this sector,[6] so powering mobility with sustainable energy is a big mission.
- **Utilities**: A sustainable energy system will be based on substituting fossil fuels with electricity from renewables. Decentralized small-scale and remote large-scale generation require a different energy transmission and distribution infrastructure; meanwhile, engaging with prosumers in new ways opens big new business opportunities.

1.8 Circular Economy

In her impressive 2015 TED talk, former professional sailor Ellen MacArthur spoke about breaking her own world record of single-handedly sailing non-stop around the globe: "Your boat is your entire world, and what you take with you when you leave is all you have," she said. "There is no 'more.'"[7]

MacArthur compares her boat to our planet, with its finite resources that we've been exploiting at alarming rates. But a sobering difference exists between a sailing boat and planet Earth: We are not heading toward a safe harbor where we can restock, grab a pint, and share our adventures at the pub.

System Earth is fairly simple: The elements on our planet have been around since Earth formed 4.5 billion years ago. For eons, our sun has been providing the energy to power wind, rain, and photosynthesis. Energy was deposited as the underground hydrocarbons we've been digging and pumping up. And that's pretty much all we have. There is no "more." (We probably shouldn't count on asteroids topping up our resources any time soon.)

Much of our modern economy is linear and single use: We take resources, make things, and dispose of them at their end of life. This results in depleting our resources and piling up waste, both on land and in the oceans. Nature is reasonably creative and successful in recycling waste: After cyanobacteria dumped huge amounts of highly toxic and corrosive oxygen into the early atmosphere, evolution invented oxygen respiration to power the metabolism of pretty much everything and everyone that walks on land, flies through the air, or swims in the water. Sometimes, even nature can't figure out what to do with waste other than piling it up—today we marvel at the cliffs of Dover, dive among coral reefs, and even live on islands formed from shells and skeletons of Cretaceous-era algae and animals.[8] Nature can even turn a waste problem into a solution.

Taking a leaf out of nature's playbook on running sustainable ecosystems should inspire us to make things from materials and then recycle these materials to make new things. This approach already works quite well with some materials like steel or aluminum, which are easy to recycle without losing quality. Organic matter decomposes naturally and goes back into short-term and long-term carbon cycles. Unfortunately, vast ranges of composite materials and plastics are difficult or even impossible to recycle efficiently: Today, these materials are burned or accumulate in landfills and the oceans.

Thought Experiment: Designing a Circular Economy

Asking where to start in designing a circular economy sounds like a strange question. (After all, one defining property of a circle is that it lacks a starting point!) We recommend looking at the different sectors of a circular economy: making, using, and recycling.

Making things begins with the design: You want to put materials in your design that can be easily reused or recycled after your product's end of life. Gluing components together is modern, but bolts or rivets are easier to undo. In manufacturing, you might put labels on components saying what they are made of, which simplifies keeping the materials cycle running. Quality design lives longer: Antiques fetch good prices, and your great-grandchildren will still enjoy them. Can we design products that live longer because they don't lose functionality and appeal, can be upgraded, or can be repaired easier?

You don't have much control over the products you sell in their use phase. But if you combine the product's sale with a service contract, or if you rent out a product or sell a product outcome while retaining product ownership, then you have full control over what happens with the product when it is returned to you. You have several options in this context: refurbish it, reuse its components, or recycle its materials. Recycling can be effective if that's what you've designed your product for.

Now, take a step back and look at the bigger picture: Many materials can be standardized, so they can be held in make-use-recycle systems and traded on markets. Do this efficiently, and you turn waste into a raw material. Incentivize customers to bring back products instead of throwing them into the landfill, which works well in some countries (like Germany) with deposits on glass, plastic bottles, and car batteries.

What are your design ideas for a circular economy?

Getting a circular economy up and running requires regulation, rules, services, and digital processes and will have at least short-term impacts on cost and prices. But the long-term reward from a circular economy will far outweigh the investment today—unless you consider it acceptable to burden future generations with our trash instead of picking up after ourselves.

Related Megatrends

- **Sustainable Energy**: Using renewable energy to power a circular economy seems like an obvious thing to do.
- **Everything as a Service**: Create customer outcomes in a continuous relationship while retaining ownership of the equipment to expand its useful life—instead of trying to sell more equipment that will end up in scrap yards or landfills.
- **Resilient Supply Networks**: Intelligent supply networks reduce waste and play a key role in closing material cycles.

Key Industries

Industries that make and sell raw materials, energy, and physical products include the following:

- **Chemicals**: The petrochemicals industry is working on reusing plastic as feedstock to substitute crude oil and natural gas.
- **Mill products**: Many metals (i.e., steel, copper, aluminum, etc.) are successfully circulated to a high degree.
- **Mining**: Coal mining is a fundamentally linear process, but mining ores is part of the ecosystem to keep the metals in circulation.
- **Building materials**: Some materials and components (like standardized steel beams) are already kept in circulation; they can be reused after the end of life of the building they have been part of.
- **Engineering, construction, and operations**: Carbon-intensive, one-way building materials are replaced with wood, and pilot projects look at modular concepts that enable the reuse of elements or materials.
- **Aerospace and defense**: The increasing use of composite materials makes recycling harder but saves weight and fuel during their lifetimes. Key components like aircraft engines are kept in service for many years, and materials like titanium and aluminum are kept in circulation.
- **Automotive**: Plastic parts carry information about the type of plastic to enable recycling instead of indiscriminate downcycling or disposal.
- **Industrial machinery and components**: Many industrial machines have a "second-life" market that keeps them in service; others are kept in ownership of the manufacturer who sells the machine outcome (product or production time) and eventually go into reuse and recycling after the machine's second end of life.

1.9 Resilient Supply Networks

Over decades, global supply chains have been optimized for maximum efficiency: Inventory was squeezed out, production was allocated to low-cost locations, materials were sourced around the globe, cheap transportation ran on tight schedules to deliver just in time. Each player made isolated, local decisions to optimize their own operations, bottom lines, and top lines. The underlying assumption of this strategy was that supply networks are elastic and will always magically adjust to all changes created by the business decisions of all players in the network.

In recent years, we've had to confront the reality that the elasticity of the global supply network to absorb shocks and disruptions is finite. Both global events (e.g., COVID-19 mitigation lockdowns) and local disruptions (e.g., a single ship clogging the Suez Canal for a week) have sent shockwaves around the globe and left the global supply network in tatters. For example, millions of cars can't be manufactured because wiring harnesses from Ukraine or chips made in Taiwan no longer arrive just in time. When an American manufacturer of baby formula decided to terminate production, a nationwide shortage ensued, resulting in the US Air Force flying baby formula in from Europe. Millions of people in the developing world are threatened by famine when ports in the Black Sea can't operate, blocked by Russia's war on Ukraine. Energy prices multiply because oil and gas are used as strategic weapons to pressure countries to engage or stay neutral, on any number of issues.

Even if you're not a supply chain aficionado, certainly you're aware that modern supply networks are complex and span the globe and that many supply chains are long and convoluted. But let's look closer at the underlying complexities:

- The topology of a supply network is generally unknown to any one participant because they only know their direct suppliers and customers.
- Suppliers don't voluntarily disclose their own suppliers to customers because they fear their customers will go straight to the source, removing the "middleman." In turn, customers don't disclose *their* customers because *they* don't want to be the middleman who is eliminated.
- Every single product has its specific supply and demand network that is meshed with the network of many other products.
- Even if the structure of the demand and supply networks for our products were shared, information about inventory levels, production backlogs and schedules, order entries, transportation schedules, and much more is generally kept under lock and key by all players in the supply network.
- Even if the supply network game is played safely, for instance, by having more than one supplier of critical components or materials, how how would you know that you're exposed to a structural supply network risk because maybe all your suppliers share a single supplier somewhere upstream? What if this upstream supplier goes out of business or faces unplanned production downtimes?

Thought Experiment: Designing a Shock-Absorbing Supply Chain

Now, you might be wondering, "How can I analyze and mitigate structural and dynamic supply and demand network risks if I have so little transparency and if I only reliably know my direct suppliers and customers?"

Imagine that you want to invite all players in a supply network for a critical product to come together in a workshop to exchange ideas about how to analyze supply networks for structural and dynamic risks. Of course, you don't know which people you want in the workshop because you mostly know your direct customers and suppliers. But if you ask your customers to extend the invite to their (business) customers, you have the demand side covered. If you ask your suppliers to extend the invite to their suppliers, then all participants in the network will eventually receive your invite—from raw materials to the distribution network for finished products.

Most participants may only agree to join a workshop if their identities aren't disclosed, so you hand out some avatars for a virtual workshop.

Workshop participants quickly agree that supply chains are vulnerable to two types of risks:

- Structural risks resulting from a player going out of business or discontinuing a product or service
- Dynamic risks resulting, for example, from transportation delays, demand spikes, insufficient safety stock, quality issues, and production outages

Perhaps your early workshop findings don't come as a surprise:

- Everybody agrees that more transparency in the supply network would enable the identification of structural and dynamic risks and provide more lead time for mitigation planning and execution.
- Nobody is willing to disclose information about suppliers and customers, which could result in competitive disadvantages.

To break this impasse, you design the following rules for a roleplay exercise to analyze the supply network for a fictitious end product:

- Players in the network can only talk to their direct suppliers and direct customers. They don't even see players outside a direct supplier/customer relationship.
- Every player sees traffic lights for the structural and dynamic risks on the demand and supply side of their product: red (risk), yellow (uncertainty), and green (no risk).
- To understand the reason for a red traffic light, players must actively collaborate with suppliers and customers to trace the issue and resolve it.

This exercise presents a design challenge: Do you need to collect all the data about the supply network in one place so the "game leader" can answer questions from the participants? Or can you keep the data decentralized, under the control of each player, and just exchange questions, answers, and reactions bilaterally?

The most resilient shock-absorbing supply chains and networks require big concepts that go beyond what individual players can do on their own. But this requirement is not a good reason to just give up and continue with business as usual. The minimum you can do is make your business as responsive as possible to supply and demand disruptions by using real-time production planning, digital sourcing and procurement, dynamic demand planning, tracking and tracing, and real-time analytics to identify issues as soon as possible and to make your response as rapid and effective as possible.

Related Megatrends

- **Sustainable Energy**: The nodes in global supply networks and the connections between them run on energy. Disruption in energy supply and erratic energy costs can send shock waves through supply networks, while an energy system based on renewable energy contributes to resilient supply chains.
- **Everything as a Service**: Delivering services on a continuous basis, instead of shipping products, strengthens the links along value chains and across industry boundaries.
- **New Customer Pathways**: Making supply networks more resilient relies on active, trusting relationships along supply chains for more transparency and responsiveness.

Key Industries

All industries that rely on physical products for their core business must have resilient supply networks for business continuity and profitable operations. This requirement obviously includes the energy and natural resources industries, and the discrete manufacturing and consumer industries. But the impact of supply network disruptions extends to related service industries in the logistics space and in financial services.

Chapter 2
Everything as a Service

A Brief History of the Services Economy

In 1995, astronomer Carl Sagan had a pessimistic view of the coming transition from the industrial economy to a service economy:

> *I have a foreboding of an America in my children's or grandchildren's time—when the United States is a service and information economy, when nearly all the manufacturing industries have slipped away to other countries...*[1]

Others predicted a nation of low-skilled workers, and similar fears arose in other developed countries. Around the same time as Sagan's words, a more optimistic expert, Gartner analyst Roy Schulte, wrote a note predicting the growth of a different breed of services in the software engineering space:

> *Developers are increasingly considering a 'service-oriented' architecture, a way of structuring programs using multitier design techniques and middleware.*[2]

Both Sagan and Schulte were prescient in their own ways. Low-wage employees in retail and food service are today some of the world's largest working populations. Meanwhile, a Google search for "service-oriented architecture" (SOA) fetches nearly 3 million references plus 16 million for "microservice," the modern form of SOA.[3]

According to the US Bureau of Labor Statistics, nearly 80% of US employment, for example, is now in the service sector.[4] Such employees are not just retail or restaurant workers but also accountants, attorneys, and architects. Something more profound has happened. Many more employees and their roles are associated with companies that have learned to create products that blend digital and labor-based services.

Jeff Immelt, former CEO of GE, co-authored an article in 2019 describing how GE's manufacturing businesses were among the earliest to morph:

> *GE was probably the first manufacturer to internalize that digital technologies could disrupt its businesses. However, that happened only after a GE*

> ***scouting team searching for megatrends serendipitously figured out through its online research that some incumbents, such as IBM, and several high-tech startups were gathering data from GE's customers to develop novel data-based services in sectors such as aviation and power.[5]***

And it's not just GE: The floodgates have opened for a new generation of offerings and business models.

Take Apple, historically known for its iconic products like the Mac and the iPhone. By 2015, iPhone sales had begun to plateau. (The 2020 launch of the iPhone 12 reversed that trend.) Apple had a platform with over a billion devices—and a loyal customer base. So, it went on a binge, generating new services and revenues from iTunes purchases, App Store fees, AppleCare warranties, Apple TV, and more. Apple now reports 825 million paying subscribers to its various services worldwide.[6] Service-related revenues have turbocharged its march to a $2 trillion stock valuation. CEO Tim Cook has disproven the naysayers who said Apple would falter after Steve Jobs passed. Cook has unleashed a new wave of innovations focused on services as much as on devices.

Conversely, for 175 years, Siemens has been known for locomotives, transformers, computer tomography and X-ray machines, wind turbines, and other complex products. But a look at its equally vibrant services portfolio today reveals an organization based on building technologies, cybersecurity, healthcare devices, consumer home appliances, and more.

E-commerce juggernaut Amazon is also a services giant in disguise. It boasts 200 million Prime members, who pay sizable subscription fees to qualify for expedited shipping, streaming music and video, and many other services. These members are increasingly buying products on a subscription basis, opting to get their toiletries or dog food delivered on set schedules. The Amazon Web Services (AWS) unit has upended the data center outsourcing services market. Amazon is a major product company in its own right. Its Lab126 has released products such as Kindle e-book readers, Fire tablets, Fire TV, Amazon Echo, and Amazon Show. Looking at this spectrum of products, one could speculate that the devices only exist to give users easier access to Amazon's media and subscription services.

For startups like Peloton, services were no mere afterthought. Its exercise bikes and treadmills are stuffed with subscription services for video streams of cycling classes, yoga sessions, strength and conditioning training, and other sessions, along with a supportive (and competitive) community of 6 million members.

Neither Sagan nor Schulte predicted this blending of labor and digital services and the creation of new bundles of physical products, software, financing/leasing, spare parts, and consumables, never mind all the after-sales monitoring, maintenance, and other services.

When we talk about "everything as a service," we're describing a trend away from the one-time or periodic sale of a product to a continuous service relationship based on the *outcomes* from a product. This transformation changes customer relationships, product design, fulfillment, maintenance, cash flows, and revenue accounting. What's more, digital technologies can become essential for delivering engaging customer experiences and making new business models profitable.

2.1 Business Model Evolution

Mark Burton could easily be a historian of the services economy. He is a Bain & Company partner focused on business-to-business (B2B) models and value-based pricing. He speaks frequently about time and materials, fixed price contracts, subscription software models, and outcome-based contracts. In our 2022 conversation, Burton reminisced about a client whose situation sounds extremely simplistic to our ears today:

> ***I worked with one of the leading IT and business process outsourcing companies. At the time, most of their business was done on a time and materials basis. They did not have incentives to deliver efficiently. If there was a cost overrun or if they misjudged the scope, they would just change the statement of work. They would throw more bodies at it and get paid more.***

Of course, their customers were unhappy with this way of doing business, so Burton was asked to come up with a model that would allow them to move from time and materials to a fixed-fee or even an outcome-based model. This project recommended a project scoping and risk assessment method and tool that allowed them to offer their services on a fixed fee basis.

Burton talked about a new generation of business models, including the following:

- Subscription: Upfront costs and usage risks for the customer are reduced.
- Usage: The customer only pays for what is consumed.
- Operational outcome: All non-financial risk is moved to the vendor.
- Financial outcome: All risk is moved to the vendor.

Burton discussed a trend that Bain & Company calls "equipment as a service" and offered a number of examples from various sectors. Munich Re's subsidiary Relayr is an Industrial Internet of Things (IIoT) startup that helps clients set up IIoT technology stacks and commercial models. Their client Aluvation has developed a micro-factory for heat-treating aluminum that fits in a shipping container. The unit can be remotely monitored and managed. Aluvation has chosen an as-a-service, throughput-based model where the company is paid by the ton of heat-treated aluminum. With remote monitoring and management, uptime has improved by 16%, and revenue growth was 40% in two and a half years.

Hilti's core business has moved from selling high-quality tools to managing fleets of tools. It has built a subscription model based on tagging, logging, and tracking pieces of highly portable equipment used on construction sites and workshops. Hilti uses customer feedback and equipment data to improve the company's model and its products. It also launched a fleet management service with outcome-based lease fees for power tools. Finally, it invested in transforming sales management and IT systems to fully support the new model. As a result, Burton said, Hilti regained market share from low-cost providers and is leading the industry in sales growth.

Burton provided additional examples and explained the business model differences between compact warehouse robotics and huge farming equipment with embedded enabling and analytics technologies:

> *One early and successful adopter of as-a-service models is warehouse robotics. You'll see pure subscription models there. You'll see outcome-based models where the customer may be charged on the number of pick-and-place actions.*

He explained that warehouse robotics share characteristics with aircraft engines because both are systems without many dependencies on other systems, and they are usually bought and maintained separately. Pick-and-place warehouse robots perform a discrete set of actions that can be managed independently from other systems, and it is easy to measure outcomes by simply counting the number of operations performed:

> *Someone like John Deere has taken a more advanced approach. They have built an ecosystem of solutions where they're not delivering everything, but they are embedding enabling technology in their equipment. External partners add analytics layers and things that allow them to maximize crop yield while minimizing inputs like fertilizer, irrigation, or crop protection. Their equipment serves as a platform and hub for the outcome-based services, so they have a really compelling productivity story to tell.*

2.2 The Trend of Servitization

Thousands of companies have attempted to follow the role models Burton has highlighted to differentiate from competitors and generate the following benefits for their business:

- More predictable revenues
- Improved attachment rates for aftermarket services

- Continuous engagement with customers, resulting in better insight into their individual experiences and needs

This trend is now dubbed *servitization*. In a world of global competition, many physical products can be copied and commoditized quickly. If you embed software to make your products smarter and add layers of innovative services, especially physically local services, you'll raise significant barriers for your competition.

Professor Tim Baines at the Aston Business School in Birmingham, UK, is executive director of its Advanced Services Group (ASG). Since 2011, ASG has focused on the adoption of servitization and outcome-based business models and is respected globally in both the research community and the industrial community for their work on servitization.

Baines cautioned that jumping from selling products to selling outcomes is rarely done in one step:

> ***When we talk about advanced services, there's an awful lot of interest now in outcome-based business models. But finding a pure version of outcome-based business model in practice is very rare. Often, they're a bit of a hybrid, so we use the term 'advanced services.'***

Over the years, ASG has developed its "services staircase," shown in Figure 2.1, that progresses from supplying products at the bottom level to guaranteeing the outcomes of a business platform at the top.

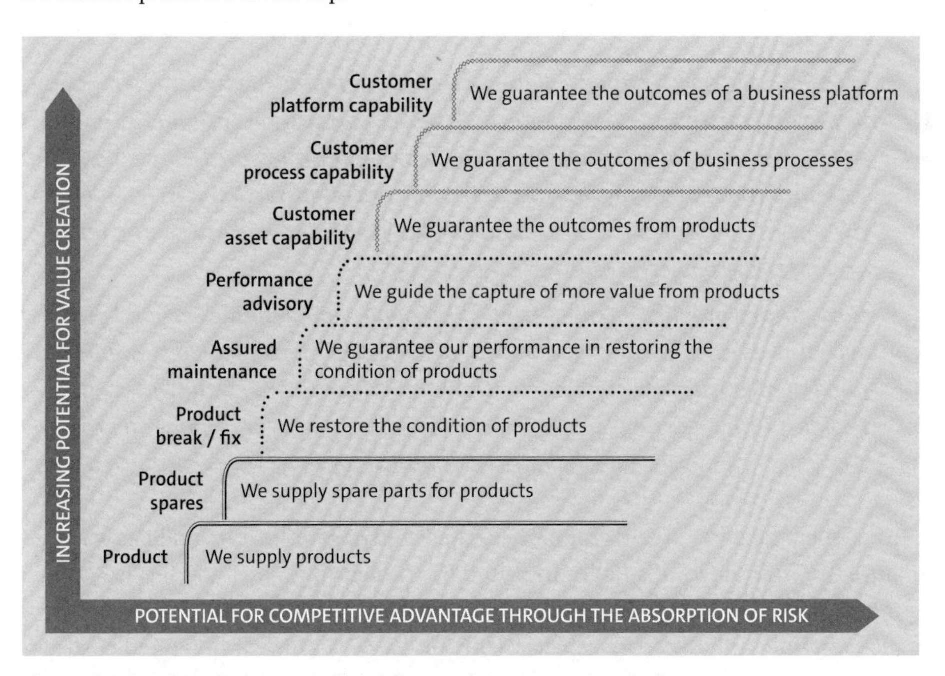

Figure 2.1 Services Staircase © The Advanced Services Group Ltd.

Baines explained that ASG started its model based on what was being used in the military under the label "performance-based logistics" and evolved this model based on work with companies like Goodyear and Rolls-Royce. He suggested looking at servitization through the lens of business model theory:

> *This perspective helps you to distinguish between what you offer the customer, how you get paid for it, and how you actually deliver it. When you go and talk to companies, the conversation about services easily derails. The staircase and business model theory helps to separate conversations about how you deliver services and how you pay for services from what you actually offer.*

Baines claimed that all services that any manufacturer could offer can be assigned to a step on the staircase, which helps with structuring business model evolution and customer engagements.

Torsten Welte wears two hats at SAP. He leads solution management for the aerospace and defense industry, looking at industry trends, focusing on enablement through leadership, and connecting development with the SAP field organization. He also leads a cross-industry team focused on the Everything as a Service megatrend, especially as complex equipment makers are learning to blend hardware, software, spare parts, and services into coherent solutions with different commercialization models.

Welte emphasized the importance of looking at things from a customer outcome perspective:

> *It's about making it easier to become a customer and stay a customer—about improving a customer's experience. Take an engineering software example: I need to design turbine blades with 3D modeling, so I want to have powerful software that makes my job easier. I don't care what version it is; I don't want to worry about software installation. I just want it to be easy to use, reliable, and offer the powerful features I need.*
>
> *Now, from a provider's point of view, this opens opportunities beyond offering software as a service. Can I bundle it with another service such as simulations? Can I add consulting services to customers' subscriptions? Can I bring another business unit or partner into play, and maybe offer 3D printing services so my customers can touch and validate their prototype?*

Welte has seen the aerospace sector pioneer the trend of servitization and has watched it mature over decades. Rolls-Royce invented the "Power-by-the-Hour" service in 1962 to support the Viper engine on the de Havilland/Hawker Siddeley 125 business jet. A complete engine and accessory replacement service was offered on a fixed-cost-per-flying-hour basis.[7] That offering has evolved into a service called TotalCare. Every time an aircraft flies, Rolls-Royce gets paid. If the offering earns revenue for the operator, it earns revenue for the original equipment manufacturer (OEM). Talk about a win-win!

In a case study, ASG wrote about the company:

> *When it comes to servitization and advanced services, Rolls-Royce—and in particular the TotalCare offering of its civil aerospace business—is one of the leading examples. Since the 1980s, Rolls-Royce has used cutting-edge technical and digital capabilities, along with engineering expertise, to provide through-life support solutions. This has been held up as one of the most significant servitization success stories, with services generating over half of the revenue of the civil aerospace business year after year.*[8]

Rolls-Royce has been diligently building a massive engine performance database for each flight. It has also created a *digital twin* of every single engine. Digital twins are models based exactly on the physical component; they use data from sensors on the physical machines to monitor operations and predict breakdowns, thus enabling maintenance that reduces downtime and ramps up reliability.

Using the volumes of rich data created by its digital twins, Rolls-Royce can reliably predict every major maintenance event an engine is likely to encounter over its lifetime. Such forecasting precision allows for planned downtimes, less frequent swapping of engines, and less unproductive time while the engine is "off wing." In an industry that is highly regulated and whose reputation rests on safety, this forecasting allows the company to, in its vernacular, "not be overly conservative."

In another example, GE Aviation provides engines on a similar "Power-by-the-Hour" business model. Its TrueChoice portfolio allows for a much wider range of services to accommodate a broad sweep of customer scenarios. Their pitch to airlines is extreme flexibility: "Fleets expand, oil prices fluctuate, operating horizons change, challenges arise. Our TrueChoice engine services suite accommodates the full range of customer needs and operational priorities with an unmatched breadth of services and materials."[9]

In the aviation sector, airlines contract for engines separately from the airframe. Airbus has a series of services branded under the Skywise family that provide health monitoring, predictive maintenance, and other services.[10]

Since the COVID-19 pandemic devastated the aviation sector, much conversation has centered on whether the evolution to services has hurt or helped companies. The ASG case study provides this perspective:

> *In 2020, Rolls-Royce hit the headlines as the global coronavirus pandemic grounded civil aircraft almost overnight. Reports that its share price had fallen to its lowest for a decade were soon followed by news of redundancies and*

re-structuring. There were comments and speculation that the main reason for the struggle was that, as a services-led company famously generating over half its revenue from services contracts linked to number flying hours, Rolls-Royce was more susceptible to the huge and sudden decrease in air traffic. In other words, it was the services business that was killing Rolls-Royce.

In actual fact, services have been the first part of the business to start to show recovery, while production rates are expected to remain at the current low levels for some time.

The ASG case study further explains the market resilience that results from a larger ratio of services to product mix:

> *The benefit to the customer is that they know their costs per hour so it makes their decision to keep operating easier as they can see whether it is profitable. Advanced services like TotalCare support the customer to be as effective as possible in their core business. Never was this more necessary than in 2020.*
>
> *The product business, by contrast, saw a greater fall in revenue because customers cancelled and deferred orders for new equipment; with aircraft grounded and no demand for increasing fleet capacity, this became a critical cost-cutting exercise for customers. Not only this, but as flying hours increase slowly, demand for manufacture will come from the services business (i.e., providing spares, not from new production).*

From SAP's perspective, servitization is widely helping to drive software upgrades. David Lowson, who runs the SAP Center of Excellence for Capgemini in Europe, has seen servitization become a top business driver for customers moving to SAP S/4HANA, noting that it has joined agility, simplification, and sustainability as motivations to upgrade software systems:

> *We have something like 470 SAP S/4HANA projects going on at the moment, and we're among the top couple of players in that market. I'm very fortunate that I get to meet a lot of pretty senior people: finance directors, CXOs, supply chain directors, and chief executives. They want to support new digital processes. The most common digital process I see is the move to everything as a service. It's almost predictable that when I talk to a large company, especially the industrial and medical ones, they will be talking about a move to providing a service rather than products.*

2.3 Cost and Revenue Modeling

Burton said one of the big lessons from moving from a product model to a service model is what he calls "swallowing the fish":

> *You have a cost curve that goes up, you have a revenue curve that goes down initially, and then comes back up. When you put the two curves together, it looks like a fish.*

Figure 2.2 shows this fish model designed by Technology & Services Industry Association.[11] Prior to an equipment-as-a-service transition, revenue exceeds costs. Then, trends reverse, and the gap widens as costs rise with new investments and revenues drop because revenue from one-time sales changes to periodic subscription revenue. Finally, the cost/revenue gap shrinks with efficiency improvements and a broadening subscription base.

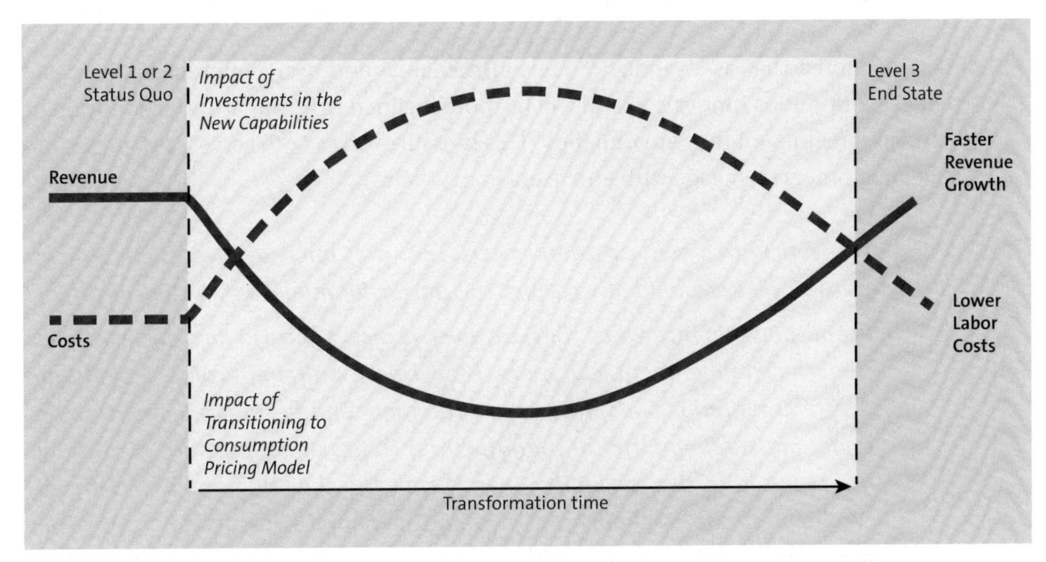

Figure 2.2 Changing Cost and Revenue Curves (Source: Technology & Services Industry Association)

If the move from a product to a service model is abrupt, the belly of the fish becomes bigger, as a one-time sale of a product with immediate revenue recognition is replaced with a service contract with periodic subscription revenues. In this transition scenario, investors need to be convinced to look at the contract value over the service term to assess performance, not just the revenue in the current quarter.

But these repercussions on financial key figures are not limited to the service providers. Organizations who become XaaS customers also need to be prepared for a shift in their financial models, balance sheet structures, and day-to-day operations. Burton had this to say:

> *Even if I am a company that wants to switch from a capital expenditure (CapEx) to an operational expenditure (OpEx) model, I have to expect ripples running through my plants. I am now dependent on an outside vendor. I'll also see impacts on my financial reporting; my numbers are going to look different.*
>
> *I've got to educate investors in terms of what our new cost model looks like, why our profit profile is changing, and what our new risk profile looks like. At the end of the day, I am probably paying more out of pocket over the life of a subscription, so I need to be darn sure that I'm getting value that exceeds that and that there's benefit to the business. It's really compelling to say I want to get out of CapEx, but the implications of doing that and the effort to do that can mean some heavy lifting.*

Lowson noted that as-a-service models also change end user preferences. They will expect continuous innovation and will be more inclined to terminate a subscription or service agreement if they are unhappy. This pressure forces both service providers and their customers to engage with end users in new ways:

> *We work with a manufacturer of cars. They presumed we always bought cars from stock. We would always pick this particular model with little variation.*
>
> *People don't do that when they are given choice, typical in an as-a-service setting. They'll experiment a bit more. They'll want a different color. They'll want a different interior. Suddenly, our procurement and fleet management comes under pressure, and the car manufacturer comes under pressure because we are not used to this flexibility. Then there's the changing end user attitude: 'Well, the car doesn't work. I'm leaving it here and I'm walking home.' They don't have a sense of ownership anymore. They feel it is the problem of the company that provides the service.*

Moving from a product-based to a service-based business model impacts several enterprise areas, from product design to balance sheet structure and customer engagement. Welte estimated that over 3,000 SAP customers across a wide range of industries could benefit from the pioneering efforts of the aerospace sector in delivering products as a service. He distinguished carefully between the opportunity for business-to-consumer (B2C) and B2B companies:

> *The major difference between B2B and B2C is the scale. If you go after B2C like GM or Apple, the number of customers you address is orders of magnitude big-*

ger than in the B2B world. The complexity, however, on the services bundling and offering on the B2B side is much higher because your customers will expect services that are tailored to their individual needs. In the B2C space, the idea of serving a 'segment of one' is still in early stages, so you can still get away with tailoring and selling products as a service to clusters of customers.

For providers, Welte said, robust market research can enable companies to design and deliver "unbeatable" product bundles to compete most effectively, allowing them to shape their markets on both the demand and the supply sides. He added that, to be successful in a B2B business, an enterprise must be agile enough to adapt rapidly to changing customer demands:

> *On the B2B side, it is a give-and-take relationship shaping your product and your solutions. It's almost like taking different building blocks, putting them together, shaping them based on the needs of your customers, tailoring to the contractual requirements that you put into place, and then delivering them consistently.*
>
> *This means that you need to be more customer-focused, and may suffer a higher cost of serving, but the cost is offset by focusing on features and options that really make the customer successful and differentiate your offering.*

Welte then described some SAP customers at various stages of their journeys to as-a-service models. For established companies, the transition to service-based offerings is gradual, taking their customer bases, their employees, and their investors along. The resulting coexistence with the traditional product business is very common. But in each industry, you'll find startups unencumbered by an existing customer base, conservative investors, or a long-standing workforce. These companies will find that starting with service-based models addresses a meaningful market segment and allows them to outpace the incumbents, who of course won't simply stand by and watch their customer base erode:

> *Companies like ABB with robotics or Tetra Pak with their packaging lines have an existing model, customer relationships, that they need to continue to support. They basically run their traditional and new models in parallel and shift over to more of an outcome-based model for specific customers.*

Welte offered several other examples from SAP customers:

- Global health technology company Philips sells a monthly subscription service to replace the heads on its Sonicare electric toothbrushes. Take this model a step further, and at some point in the future, you may start buying "dental health as a service," powered by sensor-studded, connected toothbrushes that monitor toothbrush usage and predict issues before damage is done.

- Signify, a spin-off of Philips, sells connected LED solutions, software, and services. Its light-as-a-service offering gives customers a variable bill based on the amount of lighting they consume. Of course, lighting is measured in lumens, so Signify has all the right incentives to create long-life, energy-efficient lighting solutions. This approach is quite different from the Phoebus cartel of 1925, when the industry agreed to limit the lifetime of light bulbs to 1,000 hours—an early and well-documented instance of planned obsolescence.[12]
- Kaeser Kompressoren SE, a leading maker of air compressors, offers its Sigma Air Utility. Customers pay a monthly fee for clean, dry, and energy-efficient compressed air instead of buying compressors. Consider the range of options for Kaeser: As long as the compressed air is delivered reliably and to spec, Kaeser may choose to deploy refurbished compressors, install backup equipment to safeguard service levels, and use its vast operational knowledge to deliver a great service with compelling profit margins.

We've seen many examples of servitization in large companies in the industrial, aviation, mining, and equipment fields, but service-based models also extend to the chemicals industry and their products. An example comes from Eastman Performance Films' automotive window tint and paint protection films. With an offering called Core, dealers (including those who don't use Eastman products) get precise digital patterns for cutting film. The precision of the digital product saves dealers time, reduces material cost, and delivers better results for customers.

Aldo Noseda, chief information officer (CIO) at Eastman, connects the use of precision patterns for paint protection film for vehicles with the transition from manual molds of teeth to computer-aided design and 3D printing in dentistry. According to Noseda, this new model facilitates customer-led adaptation:

> ***Instead of using a mold, dentists these days use computer-aided design, laser scanners, and 3D printers. We have similarly made the job of the dealers more efficient, and we have brought them patterns to vehicle parts to accelerate their work and reduce their cost.***

The dealer interactively creates the design in collaboration with customers. After selecting their car's make and model, a customer can preview the result and make further customizations. Finally, the system will cut out the film, reducing the opportunity for manual errors and thus resulting in a better service with less waste.

For Noseda and Eastman, the shift from products to services has expanded the company's market:

> *Our prime business is selling film, but we are also selling our software and the intellectual property of the patterns, as a service, on a subscription basis. We are up to 3,000 customers in North America, and we're expanding globally. We just launched in China, a very important market for paint protection film. We're also expanding to markets like Europe and Asia Pacific, where we hope to gain strong market share early on.*

Noseda provided another example involving thermal fluids, which have heating and cooling properties used in more than 50 industrial processes involving heat transfer. Eastman offers, among other products, a family of Therminol heat-stable fluids developed specifically for the indirect transfer of process heat. Noseda explained that heat transfer fluids must be sampled and tested at least yearly to prevent unplanned system downtimes. Eastman's Fluid Genius service offers analytics and advice regarding how long a fluid can be used and how to improve or maintain the fluid quality. The system uses proprietary mathematical models based on data from processing samples collected from customers over decades. Eastman serves more than 1,000 customer locations with Fluid Genius, based on the value proposition of optimized usage and costs for heat transfer fluids while reducing the risk of unplanned production disruption.

2.4 IT Shopping List

To deliver on the as-a-service model, industrial companies use a complex set of business and technical capabilities, starting with classical ERP functions like manufacturing, maintenance, financial accounting, or subscription billing and ending with technical capabilities including Internet of Things (IoT) sensors and the digital twins to represent physical machines.

To start compiling a digital capability shopping list, Lowson listed some early stages: Know the product you're building (including all its connectivity parameters) to govern and control the process and then configure, price, and quote the product based on market analysis and current trends. Finally, you must deliver, measure, and bill for the services delivered. With the next stage, planning, Lowson said that SAP solutions like the ones shown in Figure 2.3 can become relevant:

> *You have to plan operations pretty effectively, with demand input coming from the customer. You've then got to keep track of the machine operations and the output produced. That's where a manufacturing execution system comes into play. And of course, you need the material flow from sourcing and*

procurement to the point of operations. So far, this is mostly classical manufacturing operations. But then you have to keep track of energy and material consumption to calculate your profitability. You need to document the outcomes produced on behalf of your customer to send correct bills based on your service level agreements. That's where SAP Billing and Revenue Innovation Management (SAP BRIM) comes into play.

Lowson stated that service delivery starts well before the first machine is turned on. Customers will look at the product configuration, batch sizes, and pricing to capture the right orders. Service providers must fulfill the compliance and quality standards expected by their customers, such as batch tracing or cold-chain management for pharmaceuticals or food.

As shown in Figure 2.3, Lowson stated that Capgemini's servitization architecture has 28 distinct "moving parts." Not all parts derive from SAP, but all should be integrated to make everything run smoothly. This integration includes removing manual steps, replacing spreadsheets, and building digital interfaces to create an integrated end-to-end process.

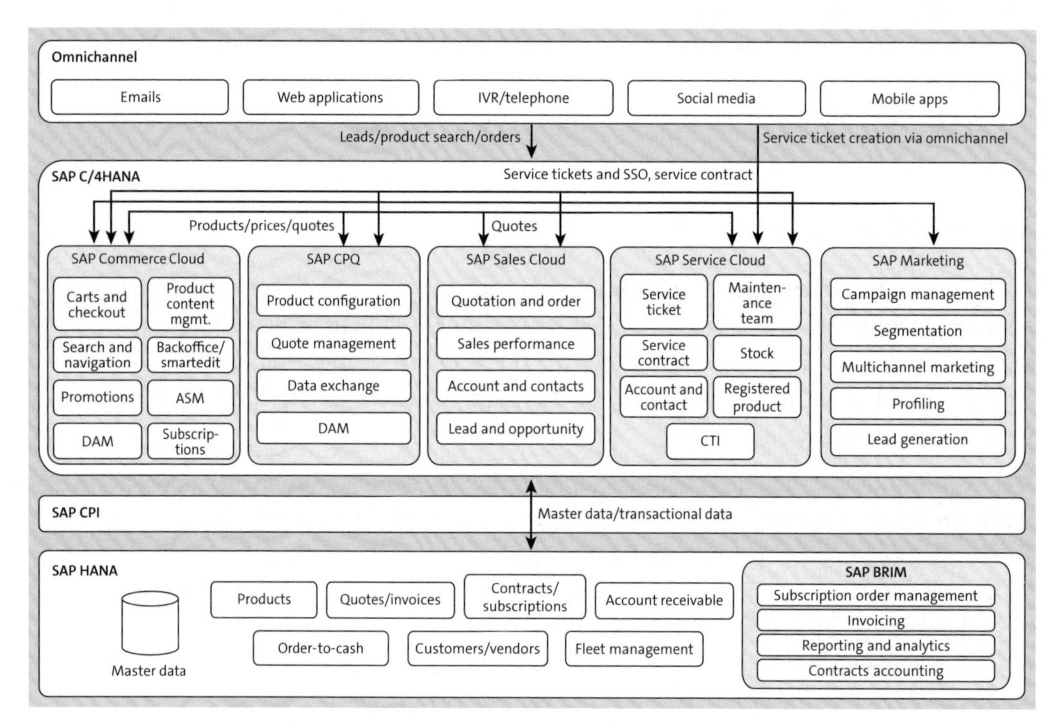

Figure 2.3 Capgemini's Servitization Architecture (Source: Capgemini)

SAP's Welte added a few key items to the technology shopping list:

> *The production assets have to become much smarter. They need embedded software, many different sensors, and they need to interpret and communicate all the data that they are capturing.*

The next step is building digital twins to capture all the data the assets are generating. Welte emphasized the importance of modeling the total operational environment of assets along their entire lifecycles. Unsurprisingly, the aerospace industry has been the trailblazer in this technology: Unplanned downtime of a machine in the workshop is a costly nuisance, but unplanned downtime of an aircraft engine can spell disaster. Welte used the example of creating an aircraft wing, from design through operations and overhaul, and described how SAP solutions can play multiple roles in the design stages and in connecting these assets:

> *SAP obviously has tools to capture manufacturing, logistics, or quality data, but we also integrate with the SAP digital intelligence layer where you can tie into other systems: how you operate the aircraft, weather information provided by the National Oceanic and Atmospheric Administration and the European Space Agency, how the stress factor was in that area, temperatures, humidity, those kinds of things. You can then enrich your digital twin with all the data, so when you next look at it from an operational perspective, you get all the surrounding data as well.*

For Welte, data acquisition and processing isn't limited to an individual asset; connecting and correlating data across a series of aircraft wings allows manufacturers to benchmark performance across the product family. This approach enables their employees and partners to access product data forms, operational manuals, and other asset information. In an as-a-service setting, a tool like SAP Business Network for Asset Management can capture IoT data to help you understand what's happening in the real world. In most contracts, operational costs are now the responsibility of the asset provider, so streamlined operations can be an important driver of profit.

In product maintenance and field service, technology adoption including mixed reality, augmented reality, and Web3 is running high to ensure service quality and profitability. Because of the prevalence of product customization in the B2B space, many product variants must be serviced, and servitization offerings are sometimes tailored to a customer segment size of "one." Cameras on devices capture data about current conditions at failure points, including images of the environment. For this category of technology use cases, mobility and cybersecurity are key.

Welte concluded with a discussion of the complex and tricky ways of owning data and information in an as-a-service context:

> *In the past, after you bought the machine, you owned the knowledge about what you produced, how much you produced, and what went into the production process. In an as-a-service setting, you want to ensure that your proprietary information does not proliferate beyond the boundaries agreed on with your service provider. Those are going to be supercritical questions that need to be answered. Who owns the data? Where does the data reside? Who is the data shared with?*

In the cloud, the question of data residency is particularly critical. Many countries have laws protecting critical public sector data and ensuring personal data privacy. But cybersecurity and data residency concerns go beyond these boundaries, according to Welte:

> *Certain countries have laws that mandate that data stays within their borders. But even if you produce and operate soda bottling machines, and if you store operational data in a country that says, 'No data can leave the country!' your data are held hostage and your business model gets stranded.*

Welte urges every company hoping to operate an as-a-service offering to ask questions about cybersecurity, data privacy, and data residency because the answers are critical for the service design: from the formulation of customer commitments and service level agreements (SLAs) to service operations and data center locations.

2.5 Organizational Turbulence

Since the list of business and technology requirements to support an as-a-service model is long, organizational alignment across hardware, software, financing, spare parts, consumables, and services can be difficult, but interesting lessons can be learned.

Bain & Company has studied various organizational models for as-a-service offerings. Burton mentioned a few key takeaways from SAP's own industry:

> *You can learn a lot of lessons from the software industry. You think back to the start of the transition from on-premise licenses to software-as-a-service (SaaS) models. The cloud natives had massive advantages because they were built around a new commercial model, and they did not have legacy systems and*

installed customer bases, and legacy ways of working that they had to gradually unwind. They were built for the new world while the incumbents had to go through a major transition with two parallel business models.

Burton speculated that the same thing will happen in the industrial space. He assumed that companies that are built or that are coming to market with a subscription model will organize themselves differently from the established players. These new companies will not have to unravel legacy organizational structures, commercial models, or ways of working. He believed that companies that are Internet 4.0 or IIoT natives will enjoy unique advantages.

However, many examples of industrial manufacturing companies in SAP's installed base have successfully managed the ramp-up of as-a-service models that complement their traditional business models. A distinct advantage is the proven foundation of SAP solutions, which can be easily complemented with solutions for delivering products as a service.

Burton explained the four main options summarized in Table 2.1 for facilitating the implementation of as-a-service models:

- In the *sandbox approach*, key units are temporarily fenced off to focus on particular solutions: "I've got an engineering group that's building hardware. I've set up a software group. I've got aftermarket service and support. I'm going to put a ringfence team across those and start iterating and sorting out what I can do and what I can bring to market." This approach is low risk, flexible, and strongly product focused—plus, as Burton noted, "It does drive cross-functional behavior, which is really key. But, at the end of the day, you're still building that on top of a diffuse decision-making structure and there can be conflicts of interest and resource availability."
- The next model is *integration into an existing service business unit*, so that only one part of the organization is affected or changed. It's a simpler way to go, but Burton pointed to the risk of parallel, potentially competing, parts of the business: one part is oriented toward the future, while the largest part of the company is operating on the established model. This parallel approach requires a lot of coordination.
- An *L-shaped organization* is Burton's favorite model. This model essentially codifies a sandbox-based approach with a solutions organization that is laser focused on delivering as-a-service or subscription models across all parts of the business, resulting in strong accountability with a clearer focus and a cross-functional view. The downside is the inevitable friction that arises as this model is put into place. Burton admitted that people's locus of control will change, but human nature tends to fight profound change.
- Finally, some organizations are establishing *separate as-a-service entities* to sidestep organizational challenges, but Burton pointed out that scalability and amplifying transformational momentum can be serious challenges.

	Sandboxes	Integration into Existing Business Units	L-Shaped Organizations	Separate Entities
Pros	▪ Low-risk launch ▪ Strong product launch	▪ Most effective for siloed functional focus ▪ Requires limited organizational changes	▪ Stronger accountability of individual solutions ▪ Ensure cross-functional view	▪ Clear accountability around resource use
Cons	▪ Decision-making effort for cross-functional topics ▪ Conflict of interest for resource use	▪ Solutions disconnected from the rest of the business	▪ Compete with existing functions	▪ Low flexibility and limited synergies ▪ Isolated from rest of business

Table 2.1 Models for Organizing an As-a-Service Business

Burton then predicted which model will prevail:

> *My view is that the L-shaped organization (or some version of that) is ultimately going to be the predominant model in the future. It just makes sense. If my business is to deliver everything as a service with all the solutions that combine hardware, software, and services, then there's really only one way to do that at the end of the day. You have to elevate decision-making authority above those siloes and then make sure that the pieces are co-developed and integrated with a service in mind coming out the back end.*

2.6 The Tetra Pak Story

Tetra Pak is a food packaging and processing company with dual head offices in Lund, Sweden, and Pully, Switzerland. The company has a proud history dating back to its founding in the early 1950s when its machines first produced small tetrahedron-shaped cream cartons. Today, it is the largest food packaging company in the world, and its impact is astounding. The company claims that, in 2021, its 54 plants produced 192 billion packages, allowing their customers to deliver 78 billion liters of food and beverages to 160 countries. In addition to selling and leasing their equipment, Tetra Pak sells packaging materials (production consumables) and services on a recurring basis directly to customers.

Alejandro Chan is the global vice president of Tetra Pak's services business. Chan described the journey to change the company's mindset to embrace the concept of service-based business models ("we make products, but we sell services"), applying the overall portfolio to provide the outcome that fits a specific customer need.

This journey started in 2012 when Tetra Pak discovered the strategic potential of expanding services as a business. "Before that, the role of services was to keep machines running and to provide spare parts and maintenance hours. We were delivering our services below the radar," Chan recalled.

> *At that time, our services, of course, generated a good level of revenues, but the emphasis was on supporting our packaging material business stream. We have since started working on expanding our services offering, focusing on long-term relationships with our customers. We are proud to have offered maintenance-as-a-service models for over a decade. Today, 40% of our services revenue comes from multiyear agreements based on a fixed price per unit produced and a performance guarantee commitment.*

Tetra Pak packages its services in three bundles—Plant Care, Plant Perform, and Plant Secure—covering proactive maintenance for assets, guaranteed performance for a full production line, and optimized plant operations for full sites, respectively.

Chan describes the company's current packaging business model as follows: A customer pays a fixed fee for every 1,000 packs produced or per liter processed, depending on the type of equipment. Tetra Pak handles maintenance, parts, and service hours to guarantee the performance of the equipment.

> *Prior to that, if the customer needed five spare parts instead of one, it was better for our services revenue because we were selling more spare parts. If the customer needed lots of support because they didn't know how to handle the machine and the spare parts themselves, it was also better for us.*

So why change this system? Chan explains that there is a limit for the total system cost to place one liter of milk in a Tetra Pak package. The price for a carton of milk on the retailer's shelf cannot go up because Tetra Pak collects increased service revenue, so their competitive position relative to packaging alternatives would suffer.

> *We started thinking about long-term service agreements where we say, 'Okay, we'll take some part of the risk. If maintenance should take 20 hours, but it takes 40 because of our fault, that's on us.' To mitigate that risk and ensure our profitability, we had to be efficient, operating a world-class supply chain to have the right spare parts at the right place at the right time. We also developed teams of competent field service engineers, deployed at a local*

level, as close to the customer as possible. This efficiency and service excellence benefits our customer's operations and also our bottom line.

Chan's focus has been to broaden the definition of services beyond today's scope to that of servitized business models, like asset-as-a-service (AaaS) or XaaS, according to customer requests. He described the growing customer expectations for different ways to consume products. But beyond the customer desire for tailored ways to do business, Chan has identified other drivers for a broader servitization agenda:

- Customers also want to lighten their balance sheets by leasing machines or by paying for machine outcomes. At the same time, opportunities arise to partner with the capital financing industry as they show a willingness to broaden their own offerings.
- Digital enablers (like sensors that allow assets to communicate their status and smart field service technologies) are cheaper and easier to deploy. This driver reduces the risk of unplanned downtimes and further increases service efficiency.
- Differentiation from the competition is required to maintain and expand the business.

As we'll discuss in Chapter 8 on the Circular Economy megatrend, acceptance has been growing of reused and refurbished technology: If the customer doesn't own the machine anyway, they will only care about quality and unit cost—not about the machine lifecycle.

Of course, not every customer immediately sees value in a packaging-as-a-service offering, especially when they realize that they need to pay for additional operational benefits and the risk transfer to Tetra Pak. As Chan explained:

> *We are talking about business value, and the value is mostly on the financial side. If you try to sell these models to the maintenance manager, or to the production manager, they may not see a big difference. To them, it is the same asset at the same spot in the factory, making the same products.*
>
> *But, if you present this model to the CEO or CFO of the company, they will consider the difference in return on capital invested. They will understand the cash flow impacts and the implications of inventory managed by the supplier. Now, why would the CFO want to talk to an industrial supplier? Because there's a lot of CapEx and working capital involved that he can free up.*

While each Tetra Pak customer will have to make their own calculations, Chan said that the company will guide them using its simulation tool to compare three business models: buy, lease, or servitize. Servitization is becoming a common customer request, he added.

Meanwhile, while evolving AaaS and XaaS models, Tetra Pak needs to adapt its back-office systems to support servitized business models. Its existing SAP ERP system was implemented 20 years ago, and the company looked at co-development with SAP using SAP S/4HANA to support innovative servitized processes. As we heard from Lowson of Capgemini, solutions like SAP Billing and Revenue Innovation Management are available to address revenue management requirements.

In summary, Chan said he has seen companies without well-established service business failing to develop successful service-based business models. These companies find that they don't have much communication with their machines, and their predictive maintenance capabilities are still immature. In contrast, he said, Tetra Pak has more than 60% of its installed bases connected to Tetra Pak to capture machine data to optimize services. And Chan was already looking ahead:

> “ *Our plan is to have a proven, commercially ready model by the end of 2023, using the current AaaS and XaaS initial cases that are already running and our experience with services agreements as the base for expansion.*

2.7 New Mindsets

As-a-service models require a mindset that few organizations inherently possess: the ability to simultaneously measure risk and reward. Burton of Bain & Company described the ideation phase where their client assesses own capabilities and their customers' unsolved problems and unmet needs. In this stage, value drivers become clear, and the value propositions that customers are willing to pay for solidify. The idea behind this new mindset seems quite simple: A customer getting more uptime on a production line because of more reliable, better-maintained equipment must have some business value. But quantifying this value and putting it into a business case can be surprisingly difficult, in particular, if you try to attach a monetary value to business risk.

Calculating the true trade-offs between value and risk when implementing outcome-based solutions, Burton said, requires a sober analysis of everything that can happen:

> “ *The naïve perception is, 'Oh, if I do everything as-a-service, I can price to value. I can price to outcomes, and I can share in that upside. I can price on the basis of operational outcomes, the number units produced, whatever.' That sounds really attractive because there's more upside, but we all know in life that with reward comes additional risk. That's absolutely true when you start thinking about outcome-based business models.*

Burton illustrated this principle with the sobering questions you should ponder when considering whether to offer anything as a service:

> “ *Imagine that I'm automating a factory. What happens if demand for my customers' products falls off? We might go through an economic contraction and I'm getting paid by the unit or percent of savings. But now the volume going through that factory is 30% lower than it was when I signed the contract. What if my customer uses old, written-off equipment for their base load and only uses my fancy 'as-a-service' offering to handle peak demand or cover breakdowns of the old machinery?*

Burton suggested that the risk minimization strategy is sometimes a simple flat-rate subscription that sacrifices upside for predictability. But another path, Burton said, involves the creation of alternative business models. He discussed demand risk and usage risk in the context of 3D printing, which offers many models. Of course, customers can go and buy a 3D printer and the consumables to supply the printer. Or they can go to a company that will set up a printer on a subscription basis. Other players consolidate demand and operate 3D printing farms: Customers send in 3D data and get finished pieces in the mail. Centralizing know-how and sharing expensive equipment is taking a page out of the cloud service provider's playbook.

Burton drove home the criticality of risk analysis because solution providers rarely have full control over the parameters that determine their profitability. Their customers continue to play a key role in as-a-service models and the associated risk can be complex to analyze and mitigate. He again pointed to the insurance sector—players like Munich Re, who are actually starting to underwrite some of that risk. This field may seem outlandish to some insurers, but on some level, it's just another risk category for which probability and impact can be quantified to calculate a premium, which is what reinsurers are amazingly good at. After all, what needs to happen is the *dimensioning of risk*, meaning that, if the risk can be quantified, it can probably be underwritten and, in turn, can smooth out the path for the creation of these models.

SAP's Welte brought up a different part of the equation: recycling and reuse. If a piece of equipment stays under the ownership of the as-a-service provider, that company has many incentives to maximize lifetimes, optimize repairability, reduce energy consumption, or extend service intervals. He illustrated this point with a familiar example:

> “ *Take home appliances. They seem to magically break down right after the warranty expires. And spare parts are either unavailable or prohibitively expensive. There is even a term for it: planned obsolescence. In the consumption societies we have lived in for the last 15 or 20 years, it didn't really matter*

what happened in the product's afterlife. But today, with scarce raw materials and a sharper focus on sustainability, the servitization mindset looks at ways to recycle, reuse, and improve your products.

For Welte, discussions about the full product lifecycle are becoming more involved these days. Years ago, the discussion about using steel versus titanium in a component was framed in a performance versus cost context. Today, the discussion spans the product lifecycle, and the thinking might go, "Titanium is much more expensive, but it's corrosion resistant and if I can actually reduce downtimes, I'd have lower maintenance costs. Plus, the titanium part may be reusable, and it definitely can be recycled, while most plastic or composite parts go to landfill." In general, economic models based on production outcomes instead of production assets tend to maximize the useful life of equipment, resulting in more sustainable and responsible operations.

Welte provided a few examples of such circular economic thinking:[13]

- Lexmark's ink cartridge collection program is designed to reuse and recycle millions of pounds of materials. Lexmark incorporates post-consumer recycled (PCR) plastic from empty ink cartridges into over 60 product components. The plastic parts that cannot be reused are converted into pellets, melted down, and molded for integration into new cartridges and printer components for potentially infinite circulation of this raw material. Lexmark's Managed Print Services program uses predictive analytics to extend the life of IT infrastructure, reducing the need to replace equipment as often as before.
- Schneider Electric, the French multinational company whose energy technologies, automation software, and services have helped companies improve efficiency and sustainability, reports that it has enabled customers to save and avoid 302 million tons of carbon dioxide emissions since 2018. Another way that Schneider Electric reduces the emission of greenhouse gases is by offering an end-of-life take-back service that recycles, reclaims, or destroys the sulfur hexafluoride (SF_6, a very powerful greenhouse gas) contained in customers' electrical equipment.
- Unilever is working with the Chilean startup Algramo to address the problem of single-use plastic pollution by allowing consumers to order household products through the Algramo app. These products are delivered straight into reusable dispensers and containers stored at home. Each refill of a product, such as laundry detergent or rice, eliminates the need to recycle plastic or send plastic waste to a landfill.
- Samsung Electronics has its Galaxy Upcycling at Home program. The initiative lets consumers extend the useful life of their outdated Galaxy smartphones by converting them into a variety of IoT devices, such as childcare monitors and smart home appliances. Consumers simply use software that repurposes the device's built-in sensors.

A segment of traditional customers still prefers to buy a product and arrange for services from a provider of their choice. In 2022, the European Commission announced the establishment of a "right to repair," including the right to repair after the legal guarantee has expired and the right for consumers to repair products themselves. Some movement in the US has occurred on this front, including many instances of state-level legislation and an executive order at the federal level. Tomás O'Leary, president of Free ICT, also advocated expanding the rights of ownership and enhancing the freedom for consumers and businesses to freely choose their providers to trade, maintain, and repair products. In a world of "everything as a service," many business models have shifted to focus on post-sales service.[14] In an interview on the Deal Architect blog, O'Leary commented, "You always have options. You've got to find a balance."[15]

As shown in Figure 2.4, ASG has developed what it calls a "transformation roadmap" that reflects the servitization journey.

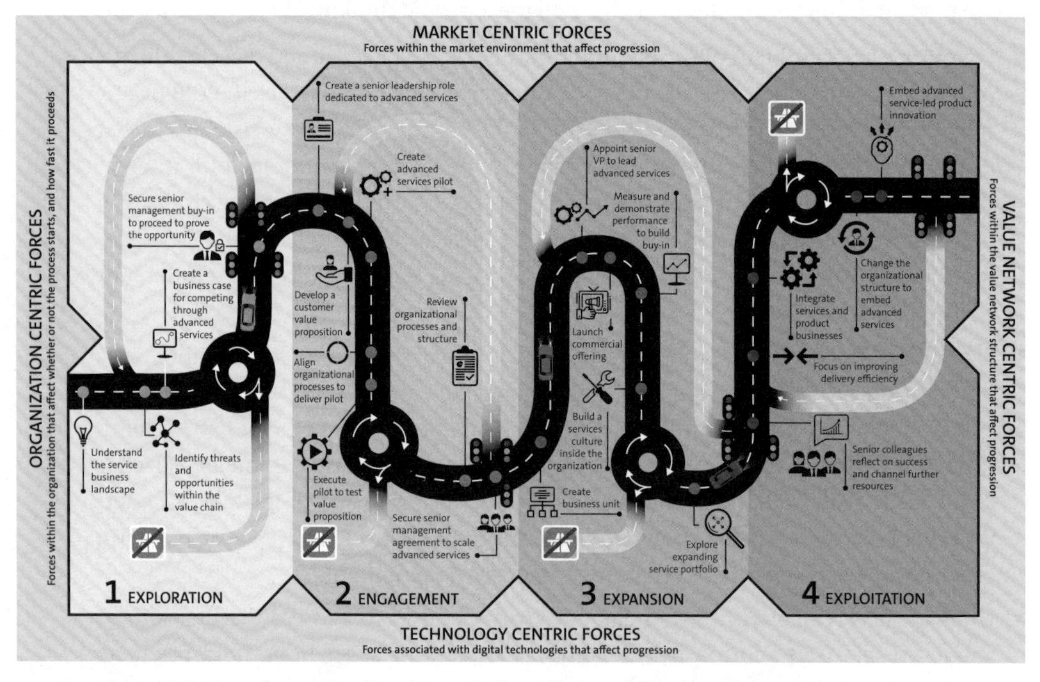

Figure 2.4 Transformation Roadmap © The Advanced Services Group Ltd.

Baines said the knowledge was developed from 30 case companies going through their own servitization journeys:

- In the *exploration stage*, customers ask, "What is servitization? Is it right for me?"
- In the *engagement phase*, they say, "Yes. I'm interested in this. I want to pilot it. I want to see if it works."

- In the *expansion phase*, they say, "Yes, it works for me. I want to scale it. I want to go across countries, across sectors, and so on."
- Finally, the *exploitation phase* is where outcome-based advanced services are in place, and now customers are looking for ways to improve efficiency and effectiveness in delivery.

In the real world, these four stages rarely happen as clean, linear steps. Companies will cycle around, for example, inside the exploration phase until they hit a trigger point that pushes them over into the engagement phase. Then, they may cycle around the engagement phase until its trigger point. Baines emphasized the concept of *punctuated equilibrium* in this model: Progress may seem to be linear from a distance, but a closer look reveals a wide range of activities happening in each of the four stages.

The dynamics of the servitization journey is strongly influenced by market forces (from both customers and competitors) and by internal forces (from the executive board all the way to the service technician). Digital technologies also play a role because they create new ways to integrate the on-site operations of equipment with a provider's control center.

What should be very clear by now is that the adoption of service-based business models is not framed only by technical parameters and customer value propositions but also by corporate culture and customer base dynamics.

SAP plays a clear role in this complex environment. We offer solutions that enable our customers to define, implement, and operate innovative as-a-service business models—from first customer touchpoints to service definition, delivery, and invoicing. As a result, SAP's customers can go through a complete business model transformation without worrying whether these new ideas can be processed by SAP systems.

2.8 Slow and Steady

Moving to an as-a-service model involves new technologies, organizational power struggles, and inching toward a new risk/reward equilibrium, so anybody who suggests this transformation is a quick fix for any business lacks expertise and experience. Regardless, we find that exploring as-a-service models is attractive and can be rewarding if done right.

Burton noted that as-a-service models have been around for quite some time and have already gone through their hype cycles. Not many integrated off-the-shelf standard solutions exist for all models, but the technology and customers' expectations are maturing. The potential to support unique and compelling models that deliver quite a lot of value is high. In his experience, many clients are actively experimenting and learning from incremental moves, so while there may not be as-a-service landslides, this approach to delivering differentiating value to customers is taking hold and is expanding.

In this growing market, slow and incremental steps make sense. Nobody can expect to develop and implement a Rolls-Royce "Power-by-the-Hour" type solution overnight. That company has evolved the offering over six decades. But organizations can begin with simpler as-a-service models. SAP is offering XaaS services to create initial momentum, kickstart planning, and overcome early obstacles.

In Welte's experience, this type of service can be a game changer:

> *We have found that many customers internally have different nomenclatures and definitions. In one of our workshops with an automotive company, we found that the heating guys, the engine guys, and the services guys did not speak a common language. We brought them together and helped them define a shared terminology. With other customers we started by talking about success definitions, and in other engagements the customer projects were advanced enough to start mapping processes to SAP capabilities and integration points.*

By now, you should clearly see how moving to as-a-service models is not a trivial undertaking because this switch affects all the functions and business units of an enterprise, from product design to sales incentives, revenue streams, and customer engagement. But SAP solutions and SAP partners are available to support the experimentation, implementation, and operations phases of delivering innovative, profitable, and value-adding as-a-service offerings to businesses. In particular, digital technologies and connectivity for monitoring and controlling equipment in the service phase has been a key enabler for innovative models. After all, it's hard to imagine how a mobility-as-a-service model—for instance, the ubiquitous electric scooter business—would work without the digital tethers tying scooters to the service provider.

2.9 Everything as a Service with SAP

A range of new or modified business and technology capabilities are required to successfully implement as-a-service models. In this section, we provide an overview of the SAP solutions that deliver these capabilities to enable the ideation, implementation, and operation of engaging, value-adding, and profitable as-a-service models.

2.9.1 Contract Relationship Management

Key performance indicators (KPIs), in particular those related to SLAs, must be captured and monitored throughout the entire customer lifecycle. Effective subscription and entitlement management is required to drive positive margins.

- SAP S/4HANA provides the customer, product, and contract master data and enables contract management with all terms for fulfillment and financial core processes.
- SAP Billing and Revenue Innovation Management enables scalable billing of simple to complex pricing structures based on metering.
- SAP Configure Price Quote enables sales executives and their support teams to tailor XaaS contracts to specific customer needs, while ensuring that products and services can be delivered and are profitably priced.
- SAP Commerce Cloud and SAP Sales Cloud enable the management of contracts and interfaces with customers as well as the management of subscriptions.
- SAP Entitlement Management controls access to digital content and manages entitlements for subscriptions by providing access to products and features for defined time periods.

2.9.2 Subscription Billing, Revenue Recognition, and Partner Settlement

Selling product usage or outcomes instead of products typically requires periodic billing based on simple subscription offerings or advanced consumption metering. Customers will also expect a single invoice integrating different as-a-service components, which is also an important prerequisite for complex revenue recognition and partner settlement processes that must be automated to enable scale and reduce cost.

The following solutions support subscription billing, revenue recognition, and partner settlement:

- SAP Billing and Revenue Innovation Management enables designing and executing revenue streams for innovative product and service combinations and subscription billing schemes.
- SAP S/4HANA Finance handles correct, efficient, and compliant revenue recognition for multiple generally accepted accounting principles (GAAP) in parallel. This solution can also automate financial processes for asset management.
- SAP Experience Management captures customer sentiment and facilitates scalable approaches to react to customer needs.

2.9.3 Digital Twins and Operational Lifecycle Management

To ensure maximum outcomes, service departments must prioritize their capabilities based on contract-related KPIs. Effective dispatch, failure identification, and issue resolution must align with uptime SLAs. Many organizations are now utilizing IoT sensors for monitoring and for applying predictive capabilities to identify repairs before issues arise.

The following solutions support the use of digital twins and operational lifecycle management:

- SAP Service Cloud supports the efficient management of service requests and facilitates parts distribution.
- SAP Asset Performance Management enables the management of assets and their operational improvement.
- SAP Business Network for Asset Management captures asset information as operated, prescribes maintenance needs, suggests actions to avoid predicted failures, and allows you to manage remote assets.
- SAP Business Technology Platform (SAP BTP) for IoT and remote monitoring is critical to capture the operation and health of assets.
- SAP Field Service Management allows for the efficient management of field services to ensure alignment with the contractual KPIs/uptimes.

2.9.4 Profitability and Risk Management

Since most services are rather customer specific, profitability management prior to drawing up contracts and during contract execution is critical. Many influential factors should be considered to manage risks.

The following solutions support profitability and risk management:

- SAP Profitability and Performance Management handles the actual performance and cost per contract and provides input for future pricing.
- SAP Analytics Cloud manages outcome-based operations that require holistic insights and reporting functionalities. SAP Analytics Cloud features an analytics platform to bring together data elements from both internal and external data sources.

2.9.5 Consumables and Spare Parts Management

Many subscription contracts cover consumables or parts. A robust supply chain that ensures the timely availability of all required parts and consumables mitigates the risk of unplanned downtimes, service level agreement (SLA) issues, revenue loss, and other potentially adverse impacts on customer satisfaction.

The following solutions support consumables and spare parts management:

- The spare parts planning functionality in SAP S/4HANA allows organizations to optimize spare parts acquisition through effective planning and distribution and by balancing inventory to address SLAs and contractual KPIs to drive profitability.

- SAP Ariba facilitates the acquisition of consumables and parts, helping ensure that the sourcing of new materials or of materials from new suppliers occurs in a timely manner.
- SAP Business Network for Supply Chain creates a network of networks to enable the sourcing and tracing of materials.

Chapter 3
Integrated Mobility

"I'm a Traveler of Both Time and Space"

Paul Salopek embodies the lyrics from Led Zeppelin's epic tune "Kashmir."[1] Since 2013, Salopek's "Out of Eden" walk for *National Geographic* is scheduled to take him more than 37,000 kilometers from Africa to the tip of South America.[2] In 2020, while navigating COVID-19 lockdowns, he wrote from Myanmar, "The ancient humans I follow hunkered for 10,000 years on the vanished land bridge between Siberia and Alaska, waiting for glaciers to melt. If nothing else, long walks teach patience."[3]

As a species, *Homo sapiens* has always been mobile. At first, we walked everywhere. Then, we learned to use beasts of burden, rafts, and boats to get around. Then, we invented the wheel, and the rest is history.

Unlike our bodies, our mobility has undergone rapid evolution. Train technology was introduced in Japan in the 1850s. Today, Tokyo's Shinjuku Station is the busiest in the world. The *Shinkansen*, Japan's "bullet train," is one of the fastest and safest in the world. The US, which saw rail track explode from 60,000 kilometers in 1865 to 400,000 in the next 50 years, has in contrast seen steep declines in passenger rail. Affordable aviation and a sprawling highway system have changed US travel habits. Mimicking both Japan and the US, China has added on average 10 airports a year in the last few decades. From zero high-speed rails in the year 2000, it now has over 40,000 kilometers of it.[4]

Other countries have evolved differently. Netherlands is known for its bicycle culture, and Indonesia for its motorbikes. Norway has the highest electric vehicle adoption rate in the world. In Saudi Arabia, nomads can be spotted ferrying camels with pickup trucks.

By the time Salopek finishes his journey, he will have witnessed the extremely wide spectrum of mobility available to modern humans. New choices keep growing as old ones persist—it's *and*, not *or*.

3.1 Not Your Dad's Auto Industry

In 2018, the German automotive conglomerate Volkswagen announced talks to form a joint venture with Didi Chuxing, China's biggest ride-hailing service. The partnership would include Volkswagen managing a fleet of around 100,000 cars for Didi, some two-thirds of which would be Volkswagen models.[5]

At first glance, this announcement might sound like a major fleet sale for Volkswagen. But Marcus Willand, a partner at MHP, a consulting firm that is part of Porsche (itself part of Volkswagen), saw it differently: "Actually, you could read this transaction as a signal that Volkswagen was slipping back one link in the mobility supply chain, going from major OEM [original equipment manufacturer] to tier-1 supplier. Half the Chinese population is on the Didi platform, and that power allows Didi to choose any OEM going forward."

Willand expanded on the nature of the cutting edge in mobility—the movement of people and goods—and the role MHP is playing in this space:

> *We are 100% focused on automotive and mobility in Germany, the US, and China, so we have a clear industry and geographical focus. This gives us the opportunity with our MHP team, comprised of 3,500 people, to give impulses to the transformation of the automotive industry. I came here six years ago to help MHP to grow from a purely automotive consulting company to a mobility consulting company.*

He emphasized this essential shift from car makers to mobility technology companies and the fact that some enterprises (especially Chinese) are leaving the old guard in the dust after Elon Musk started this transition at Tesla:

> *It's not a car, it's a computer. I'm plugging four wheels to it, and building a body around it, and then someone can drive in it—but, in the end, it's a computer.*

The IT architecture that Tesla built into the Model S was fundamentally different from any platform architecture of competing OEMs, and most of those OEMs are still struggling to catch up. German engineers may be heartbroken to hear Willand's explanation that the new Chinese companies would not copy German designs when building cars, but would instead imitate Tesla's approach and try to do even better. Willand had the opportunity to work with Chinese electric vehicle producer Xiaopeng Motors (commonly known as XPeng). In the first half of 2022, it sold more electric vehicles in China than Volkswagen. XPeng was founded in 2010, so it's much younger than Tesla, but Willand's articulation of its credo should sound familiar:

> *They say, 'We are building a smart device. Our focus is autonomous driving software and car entertainment systems, so we are basically a software provider.'*

In comparison, Willand says, the traditional automotive industry seems to be stuck in their drivetrains and is years behind producing state-of-the-art software. Players like Waymo on the US side and Didi in China threaten to capture large portions of the mobility market.

Hagen Heubach, global vice president of SAP's automotive industry business unit, is responsible for leading the entire end-to-end solution portfolio for automotive customers at SAP. His department is responsible for formulating a central strategy and pushing strategic engagements in the automotive space and the future of mobility. Through his customer engagements, he has observed a profound change in attitude:

> *All our customers—OEMs, truck manufacturers, bike manufacturers, suppliers, national sales companies, mobility providers—are facing a run-and-disrupt scenario. They say 'Hey, I need to run a very profitable and traditional business, and, at the same time, I need to disrupt myself into this new, sustainable mobility world.'*

Heubach and his team have identified six strategic priorities for automotive OEMs and their suppliers who are long-term SAP customers:

- **Customer-centricity**
  "The first is customer interaction. The first week of COVID, we started to get calls from several OEMs saying, 'We need to change to direct digital sales because all the dealerships are closed.' The new Ford Mustang SUV—the electric Mach-E—is likely to only be sold through direct digital channels. We've seen a rapid disintermediation: the OEMs are taking out the middleman (the car dealership) and establishing a direct relationship with the driver, something that we have been discussing many years ad nauseam."
- **Anything as a service**
  "The second priority is around new business models, including the servitization trend. Not just the OEMs, but also the component manufacturers and automotive suppliers are looking into new business models bundling products, services, and software to monetize the packages on an as-a-service basis."
- **Connected smart products**
  "Third is connecting the car and treating the car and its components as a connected platform. Think of the data we are collecting around predictive maintenance, customer experience, and other areas. Live vehicle and driver data promise to unlock important new revenue streams."
- **Resilient supply chain and smart factory**
  "Fourth, the mobility supply chain needs to be 100% resilient. You don't want too many issues like many German OEMs had with wiring harness suppliers during the Ukraine crisis."

- **Sustainability and circular economy**
  "Fifth, you will not survive in the mobility market if you cannot prove that your components and materials are sustainably sourced—in particular in the e-mobility space. Conventional cars made from steel and aluminum are reasonably circular and we have to apply the same approach to batteries, electronics, and advanced composite materials."
- **Workforce transformation**
  "Finally, there is another one strategic priority we discuss with several customers: transforming their workforce. All our customers are moving from classical engineering to software company cultures."

The strategic priorities of the automotive and mobility industry need to be digitally enabled and align nicely with SAP's solution portfolio and roadmap.

3.2 The Case for CASE

In the last few years, CASE has become a rallying cry for rapid change in the automotive sector.[6] In this framework, vehicles and mobility are *connected, autonomous, shared,* and *electric*. In the discussions for this book, both Willand and Heubach made the case for CASE.

3.2.1 Connected Cars

Willand portrayed BMW as a long-time leader in connectivity, and BMW itself highlights some nice historical trivia on its website.[7] In 1980, BMW integrated the onboard computer in its Formula 1 racecars. In 1996, it produced the first cars to be equipped with an emergency call function. From 2000, GPS took navigation to a new level and has also allowed stolen vehicles to be traced. In 2007, the introduction of the iPhone turbocharged the category of auto infotainment. Since 2010, cars have been aware of the environment and are now capable of exchanging information with other vehicles, with the cloud, and even with pedestrians. *Vehicle-to-everything* (V2X) communication has become more common.

Regarding the opportunity for communication between vehicles and traffic infrastructure, Willand sees "traffic lights or sensors that guide cars, so that the infrastructure takes over assistance and some responsibility for autonomous driving."

3.2.2 Autonomous Cars

The industry has settled on a taxonomy for levels of driverless autonomy based on how much the driver or the vehicle is in charge: Level 0 (no automation), Level 1 (driver assistance), Level 2 (partial automation), Level 3 (conditional automation), Level 4 (high automation), and Level 5 (full automation). Level 0 was the baseline of 30 or 40 years ago when the Defense Advanced Research Projects Agency (DARPA) in the US encouraged young innovators to design versions of driverless cars. Willand commented on the speed at which technology is evolving:

> *XPeng has managed to get their latest product, the P7, which is the equivalent of the Mercedes EQS, the S-class, to autonomy level 3. In their first attempt, they managed to get that far. In some functions, they're even better than Tesla. It's really impressive how fast they have advanced.*

Willand said he thinks the industry can reach Level 4 or even Level 4+ by the end of this decade. In terms of usage, however, he saw the most promising use cases in the trucking sector:

> *We stated five years ago: 'The biggest business case for autonomous driving is in the logistics industry.' Platooning of trucks with a human-driven lead followed by a train of autonomous trucks on designated highway lanes appear to be the most promising. That would be especially true for markets like the US with consistent truck driver shortages.*

He thought individual mobility was a little trickier. In his view, the use of automated vehicles might work with dedicated lanes and disciplined drivers in certain settings:

> *Driving in downtown New York during rush hour might work because the traffic moves slowly, and drivers mostly follow the rules. But look at Delhi, Cairo, or Naples (Italy) where the traffic is much more chaotic, I'm not so sure. So much fuzzy logic is needed. It's tough to imagine an AI system capable of learning so many exceptions.*

3.2.3 Shared Mobility

For Willand, the new emphasis on sharing transportation through ride sharing, trains, buses, carpooling, and so on is shaking up the industry in many ways. He stated that the Didi approach is a precursor to a broader trend. If cars are sold in fleets, car dealerships become obsolete. For Volkswagen, fleet sales are booming, and deal sizes are getting larger. He expected that the mobility of the future will be mostly a fleet-driven business, which is bad news for American car nuts, British petrolheads, and German *autoliebhaber*s.

Heubach added:

> *Many OEMs are experienced in selling fleets to rental car companies. Volkswagen even acquired Europcar. What's new: now they are looking to apply anything-as-a-service concepts as in the aerospace sector. They apply concepts like digital*

twins to better manage the fleets in the streets, essential if the commercial model moves to usage-based billing.

Heubach also pointed out that, in Europe especially, where company cars are often part of compensation packages, corporations themselves are the biggest buyers and lessees of fleets. Thus, fleets have become even more of a business-to-business (B2B) industry.

Car-sharing models have emerged in large cities where owning a car can be more of a burden than a liberation. But take a stroll through Berlin or Rome, or across the SAP campus in Walldorf, and it looks like shared mobility accelerates on two wheels: Hopping on a bicycle or grabbing one of the ubiquitous e-scooters to zip from the train station to the office or to the restaurant is super-convenient and has rapidly become normal. It stands to reason that these mobility devices may become another disruption, from below: People who routinely use shared two-wheelers will quickly adopt a four-wheeler if they offer a compelling mobility experience.

3.2.4 Powering E-Mobility

Finally, Willand focused on the electrification element and described the "double-whammy" of the move to electric vehicles—a dramatic simplification of the bill of materials (BOM) and the growing role of in-car software:

> *The biggest barrier to enter the traditional automotive industry was the complex drivetrain. The transmission and the engine were the most sophisticated parts of the cars. This barrier basically went away because the electric drivetrain is almost depressingly simple. Most electric cars have a single-speed direct-drive gearbox and instead of shuffling gears around to go in reverse, a switch lets the motor rotate in the other direction.*

This simplicity created room for new OEMs who would buy most components off the shelf. XPeng even outsourced the bodywork; it is just buying and assembling components. Willand argues that the differentiator in e-mobility is the software, an area where traditional car makers are still struggling. These traditional companies are facing two threats at once because all their drivetrain know-how accumulated over decades is depreciating rapidly and because looking for new areas of competitive differentiation is still in an uphill battle.

But how about the batteries? Willand said that batteries are in the first phase of industrialization. He doesn't think that we can expect that Moore's Law of exponential improvements applies. But he argues that even linear improvements of capacity per unit of weight are important steps to further accelerate the transition to electric mobility.

Batteries and electric motors also raise sustainability issues. They currently require lithium, cobalt, nickel, and other minerals that are in short supply and will be subject to guidelines for sustainable and ethical sourcing. A circular economy approach is required to recycle battery components. Ford has announced that customers leasing any of the company's electric vehicles no longer have the option of buying out the vehicle at the end of the lease.[8] This policy allows Ford to recycle the batteries instead of continuously making or sourcing new batteries.

Regardless of battery materials and business models, batteries need to be charged to keep the car moving, so Willand added:

> *We are working with SAP to make the battery charging experience seamless. That means, no matter who the charging provider is or what charging port you are driving to, you always have the same experience. All the authorization, metering, and invoicing needs to be done seamlessly, in the background.*

This promise is music to the ears of early adopters who have had to carry a stack of charge point cards when they drive across Europe or even just into the territory of a neighboring charge point operator. But Willand's vision goes beyond a hassle-free charging experience:

> *Down the line, we may also see some really interesting scenarios. In Germany, a power plant is defined as a facility that can deliver at least one megawatt. Combine seven Porsche Taycans and you have a power plant that can push electricity from the Taycan battery packs to the grid. You could sell it when the price is highest, and then recharge the cars shortly before you drive—at possibly a cheaper rate. This arbitrage business is interesting if you scale it to millions of electric vehicles. This approach will create a market that is called 'vehicle to grid' or V2G.*

Willand pointed out that not every economy on this planet runs on cars—he mentioned electric tuk-tuks in Indonesia hired via digital platforms—and that not every company can simultaneously handle the change across the CASE paradigm. In his view, only a handful of OEMs, such as Toyota and Volkswagen, can implement all four elements of CASE under their own roof. Others will focus more on autonomy and electrification alone. To support this assumption, Heubach pointed out that Daimler and BMW are selling Share Now, their ride-sharing joint venture, to Stellantis.

Even if the car manufacturers carefully choose a focus in the CASE space, they must still keep up with evolving regulations and with changes in consumer demand. Electric vehicle sales represented just 9% of all passenger car sales in 2021, so even with rapidly increasing market share, especially in China and Europe, there's a long way to go for e-mobility.[9] Stellantis CEO Carlos Tavares has lamented that policymakers, in their publicly announced timelines, appear to "not care" whether automakers have enough raw materials (or time) to support the shift to electric vehicles.[10] ExxonMobil CEO

Darren Woods said in an interview that their internal scenarios show that every car in the world will be electric by 2040, interestingly without being concerned that Exxon-Mobil's overall business would suffer significantly.[11] We're a still long way off from this scenario.

One consequence of this slow-motion shift in mobility is that SAP has had to support automotive customers and their business models in multiple dimensions as they gradually transition. Given the need to support both "run and disrupt" models, Heubach said that SAP has taken a two-tier approach: support for traditional industrial approaches and support for emerging digital approaches, as shown in Figure 3.1.

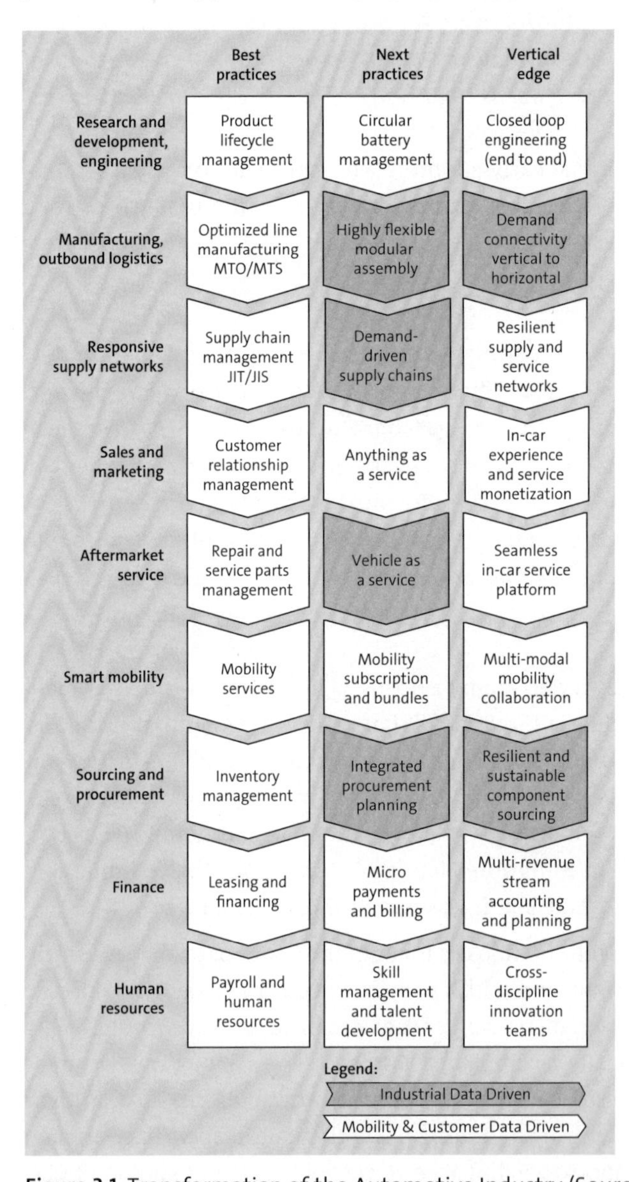

Figure 3.1 Transformation of the Automotive Industry (Source: SAP)

For Heubach, blending industrial and digital mindsets has led to greater expectations for software supporting the automotive industry. He mentioned some new features that you would not typically expect from SAP—including SAP E-Mobility, a solution for charge point management and operations. He also focuses on helping OEMs develop new sales channels, make supply networks more resilient, and improve emission management efficiency:

> *Even one step further, and this is very important as to where we're going in business networks. Not just the traditional OEMs and their tiers of suppliers. All the challenges—what we're seeing about the chip shortage, the changing world of sustainable coatings and chemicals, regulations around scope 3 emissions, etc.—are driving better visibility across networks.*

He then spoke about SAP's support in three key areas:

- **Industry best practices**
  It is easy to forget the auto industry has been evolving since Karl Benz invented an internal combustion engine and built the first passenger car, Henry Ford installed the first assembly line, and the Japanese brought in continuous improvement mindsets and lean manufacturing principles (*kaizen*). Heubach said practices and scenarios such as Kanban, just in time, just in sequence, make to order, and make to stock have long been accepted as best practices and have been supported by SAP's software for many years. On this foundation, today there is a lot more Industry 4.0-driven automation on the shop floor and in the warehouse that the industry has been incorporating.
- **Next practices**
  For Heubach, the automotive industry needs new capabilities across many business areas to succeed in the "run and disrupt" game. He mentions visibility in the supply chain and support for new business models like pay-per-use or subscription management. Some of these practices look familiar and may be already supported by SAP solutions, but transplanting practices across industry boundaries is never a trivial undertaking.
- **Vertical edge**
  Here, Heubach took us into a more experimental space that leaves the firm ground of well-established best practices and the proven need for next practices. He outlined how SAP S/4HANA, SAP Business Technology Platform (SAP BTP), and SAP's industry cloud are the right environment to experiment with innovative processes, intelligent technologies, and digital support for new business ideas. For Heubach, these solutions open up the innovation space: "We are partnering with our customers, with hyperscalers, system integrators, industry experts, and other solution providers. The next in-car software probably won't come from our software labs, but we and our partners will play a key role in the mobility stack."

Making electric vehicles move needs powerful power packs—and this is an innovation space way beyond the automotive industry. As shown in Figure 3.2, making batteries and pushing their performance to new levels involves a wide range of industries such as mining, chemicals, utilities, mill products, and high tech involved in the lifecycle of developing, producing, servicing, and recycling batteries for electric vehicles.

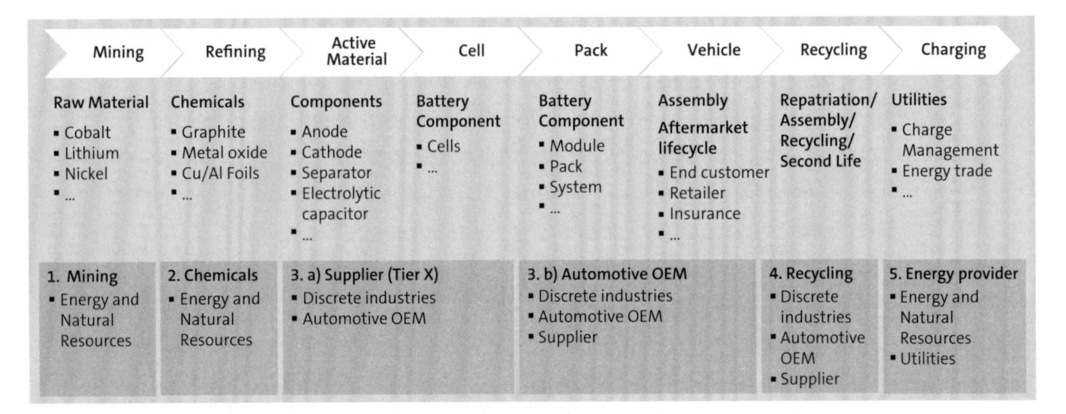

Figure 3.2 Battery Lifecycle Management (Source: SAP)

3.3 E-Mobility

The European Union (EU) has been championing the global adoption of electric vehicles. The publication *Recharge EU*,[12] by the research organization Transport & Environment,[13] laid out the challenge for the charging infrastructure: "To keep up with this electric surge—estimated to be between 33 million electric cars in the current policy scenario and 44 million in the climate neutral one in 2030—EU's infrastructure framework needs to prioritize electric charging and be in line with the increasing demand for public and private charge points."

The report outlines policies and projects, such as the following:

- At commercial properties, one-fifth of parking spots should be outfitted with chargers.
- Public funding should be used to upgrade electrical grids in cities.
- Charging hubs should be available in cities, with priority given to taxis and shared vehicles.
- Intercity road networks should facilitate vehicle charging even in remote areas.

The report differentiates between charging scenarios: charging at home, at work, and on the road, both close to home and long distance. Focus should center on access to charging stations. In 2014, an EU-appointed commission suggested a maximum of 10 electric vehicles per public charging point (PCP) across Europe to ensure enough room for everyone to charge when needed.

This infrastructure is quite different from the gas station network we are familiar with. Conventional cars refuel in a few minutes, and "range anxiety" is not an issue when the next gas station is around the corner. Compare this convenience to the experience of e-mobility pioneers who spent hours at a charge point (if they found one!) and who anxiously watched the battery gauge instead of enjoying the scenery.

Battery capacity and range have grown, and charging times have shrunk, but the e-mobility infrastructure will continue to be highly decentralized. SAP sees business opportunity in the emerging network of charging points. Jörg Ferchow, chief solution manager at SAP, provides an overview of SAP E-Mobility to build, run, and manage such networks. In this model, vendor-independent charging devices would connect to a cloud solution that uses the Open Charge Point Protocol (OCPP) that has become predominant in the market.

In this young but fast-growing market, SAP expects a wide range of partners who build, service, resell, and operate charge points. Even though electric vehicles only make up about 5% of total vehicles on the road in 2022, already a complex ecosystem of players and processes must be navigated and integrated. For Ferchow, SAP can bring some level of discipline for improved overall efficiency and a more convenient and seamless customer experience:

> *You have the e-mobility service providers. There are charge point operators. If you travel long distance, you may have to deal with roaming providers, as we have with mobile telephones. These providers have contracts with many charge point operators. As a driver you need a charging box in your garage or apartment parking space. Of course, these days we are seeing many more 'prosumers' who don't just buy electricity, they sell power back to the grid. They may have solar panels and store surplus energy in the car battery and at times sell it back to the grid. Or they may be part of a growing number of virtual power plants, which share energy across communities of consumers.*

SAP envisions simplicity and convenience for the driver—plug in and that's it. We compare the current e-mobility infrastructure with the early telephone network with its switchboards, complicated long-distance calls, coin-operated phone booths, and scheduled international calls. Today, all this complexity is hidden: You dial a number, and the telephone rings a few seconds later, regardless where on the planet you and the person you're calling are located. For e-mobility, there is still a way to go.

Ferchow explained the SAP E-Mobility solution, shown in Figure 3.3:

- "In the charge-at-home scenario, you're dealing with company cars. You charge your car at home, and your company reimburses you. That's pretty simple: just connect to SAP Concur, our spend management solution."

- A charge-at-work scenario uses SAP E-Mobility, SAP S/4HANA Cloud, and the SAP digital payments add-on. "This scenario is aimed at companies with fleets of electric vehicles. Some of them even make their stations available to the public outside of office hours and the charge points will show up in a roaming provider's app," he said.
- The charge-in-public scenario relies on SAP Subscription Billing and industry cloud solutions from partners. "For this scenario we expect to see a number of oil and gas companies," said Ferchow. "They are putting charging infrastructure at gas stations. Their retail stores also want to attract customers, and they want to combine the charging experience with the shopping experience." Fast-charging a car may take 20 or 30 minutes, well over the 3 minutes it takes to top off a gas tank. Just imagine you're at a gas station and your car needs 30 minutes to charge, but you have a couple of restless kids with you. You're a mobility hostage!

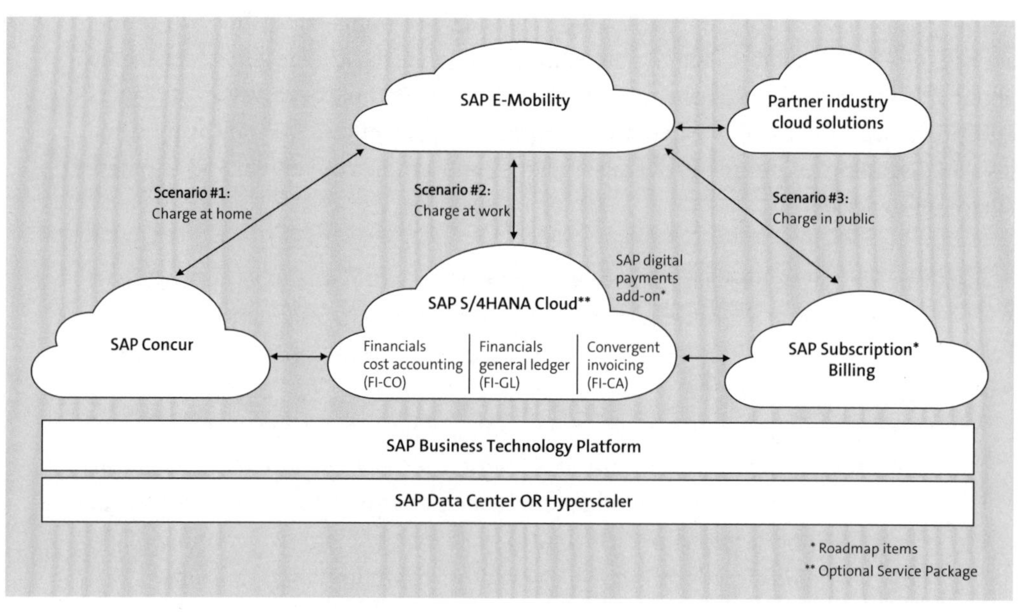

Figure 3.3 Architecture Overview of SAP E-Mobility (Source: SAP)

Steffen Krautwasser manages SAP's impressive global vehicle fleet of 27,000 cars, 17,000 of which are in Germany. He explained how the fleet has evolved and how he has been one of the early adopters of the SAP E-Mobility solution:

> *As you know, Germans have a love affair with their cars. When SAP was a much younger company and Walldorf was a smaller village, a company car as a perk turned into a recruiting advantage to attract candidates from larger cities like Munich and Berlin. While our car policies vary from country to country, every employee in Germany who has at least three years of service qualifies for a company car. SAP covers most of the cost: the maintenance, the fuel,*

the insurance. The employee gets a new car every four years. It is a really nice perk.

As SAP has grown, so has the size of its fleet; meanwhile, running a sustainable company has become more and more important for SAP's investors, for SAP's employees, and for the community. Krautwasser discussed how the makeup of the SAP fleet aligns with SAP's corporate sustainability strategy:

> “ *Almost a decade ago, we started introducing battery electric vehicles (BEVs) to our fleet. We were one of the earliest companies to adopt them. Initially they were on a short leash, mostly driving around on our campus. Today, almost 20% of our fleet is electrified vehicles. The definition includes BEVs and plug-in hybrids. But starting in 2025, every new car our employees can order will be a BEV. Hydrogen-powered vehicles do not look like an alternative for now. It may change, but for now we are focused on BEVs. We plan to have no cars with internal combustion engines by 2030. This is realistic because we return the cars after four years.*

Going electric comes with additional complications in delivering this mobility incentive for employees:

> “ *We give our employees gas cards they can use at most gas stations. That doesn't work for electric cars. The network of public charging stations is getting better every day, but it still has a long way to go. So, we had to think about charging at SAP locations, and about charging at home. At Walldorf, our headquarters, we are already operating 600 charge points.*

For a single corporate campus, 600 charge points is a quite significant number. Krautwasser assumes he will need 1 charging station for every 5 cars, which means that he has to plan for a setup of 3,500 charging points in 2030. Already this estimate has required elaborate planning and coordination between SAP's facilities group and its local electric utility. Bolting charge points to garage walls is the smallest issue. Getting the necessary power infrastructure up and running requires detailed planning and sophisticated agreements with utility providers. Krautwasser is starting these conversations early to maximize the lead time for utility providers to dig trenches for thicker cables and install transformers capable of handling the expected loads. He is also aware that distribution system operators (DSOs), the utilities in charge of managing the electricity distribution from the substations to each meter, are struggling to power up the grid to handle additional demand from the proliferation of electric vehicles.

Supporting employees to charge their vehicles at home can significantly reduce the strain on the infrastructure on SAP campuses, according to Krautwasser:

> *We reimburse for installation of the wall charging unit and for the power consumption. During COVID, clearly that was very convenient as most of our employees worked from home. But even as people come back to the office, they still have a dedicated charging station at home and don't have to worry about finding one at the office. In some locations outside of Germany, the expenses are reimbursed via SAP Concur in a fully automated digital no-touch process.*

Krautwasser noted that the operational cost to SAP of charging at the office is lower because of the favorable large-customer rate SAP enjoys, but the capital costs of the campus infrastructure are much higher. From SAP's perspective, it is a bit of a wash whether employees charge at home or at the office.

The longstanding company car perk may be less desirable for younger generations of SAP employees who tend to live in urban areas and bigger cities and thus can comfortably rely on public transport or an occasional rental car. Heubach provided an interesting perspective of his younger employees:

> *They tell our fleet manager, 'I am not interested in a company car. I'd rather have a mobility budget from SAP, and I can decide how to use it: do I want my own dedicated car, do I want to subscribe to a ride-sharing system, or do I want to have micro-mobility with e-scooters combined with public transportation.'*

To provide his new colleagues with alternative mobility perks, Krautwasser says a new model is almost ready for rollout:

> *We are piloting a mobility allowance. It's actually more of a virtual budget. You can go and buy public transportation tickets. You can spend it on car sharing or bike sharing. You can use any kind of mobility, and you'll get reimbursed via SAP Concur up to your approved budget. The overall feedback from employees, after two and a half years of the pilot, is overwhelmingly positive. The feedback is 'Hey, you're taking the topic of sustainable and integrated mobility really seriously.'*

Of course, these transportation options aren't limited to the younger SAP employees. Krautwasser readily acknowledges that he prefers to take trains on business trips. For him and his colleagues, integrated mobility is a world of *and*, not *or*.

3.4 Transportation Sharing

Shared mobility is definitely not a new concept. In some communities, geographic and spatial limitations, coupled with population trends, have resulted in a long history of shared mobility.

Augsburg in Bavaria is a university and tourist town an hour's drive west of Munich. Augsburg is famous for the Fuggerei, the world's oldest public housing complex still in use, founded in 1521 by the rich Fugger family of merchants.

The municipal utility company Stadtwerke Augsburg (swa) supplies electricity, natural gas, district heating (the next generation of "steam heating"), and drinking water for 350,000 people, and it has also expanded into the mobility space, pioneering a flat rate mobility package in 2019.[14] Isabella von Aspern leads digital transformation at swa and spoke about matching mobility offerings with the services demanded by citizens in her community:

> *We are a very green city surrounded by lakes and forests. Being so close to BMW headquarters, of course we have a car culture. But among the younger generation, especially in cities with good public transport infrastructure, the desire to own a car is declining.*

To meet citizens' demand for transportation, swa offers fixed-price mobility plans that allow customers to take advantage of all modes of transportation, such as buses, trams, and rental bikes. The metered cost-sharing component registers use by time and distance; people who exceed the allotted time or distance simply pay for additional use.

von Aspern drew attention to one major component of swa's services:

> *Car-sharing has become especially popular. It has really grown over the years. We started with a few cars and a few places where you could pick them up. Our customers did not expect it from us. It was service that our CEO believed would be appreciated. He is very innovative. The big difference from other models: we run our own operations and perform all the service and maintenance.*

Outside mobility, other swa services have been well received by its customers, von Aspern said. She feels the utility is an important contributor to the health and well-being of the Augsburg community:

> *People are very, very happy with our water. It is super safe for drinking. But they also tell us our buses and trams are very clean. We receive many compliments about this. Additionally, our stops have surveillance cameras and are safe. I think the city feels pretty safe and clean, and I think we are, as Stadtwerke Augsburg, a very big part of it.*

3.5 Mobility as a Service

Kevin Schock, vice president of the travel and transportation industries at SAP, has extensive experience in the aviation and railroad sectors, and his definition of mobility encompasses many modes of moving people and goods. Within his purview are customers such as Amtrak, Deutsche Bahn, and JR East-West, the last of which runs the Japanese bullet train, the *Shinkansen*. Also included in Schock's job are solutions for supply chain resilience, to which the movement of goods in a complex global system is central. For example, goods may arrive in the Port of Los Angeles on a vessel operated by French container shipping company CMA CGM France and then move to Chicago on a major US railroad, such as BNSF Railway.

While Schock focused on long-distance mobility, his colleague Senta Belay is more concerned with urban mobility, for which the single-occupancy automobile has long created the biggest headaches. To illustrate the point, Belay referenced an iconic poster created in 1991 by the city of Münster (dubbed Germany's bicycle capital) to encourage bus usage by juxtaposing the relative space taken up by 72 bicycles, 72 cars, and 72 passengers on a bus, as shown in Figure 3.4.

Figure 3.4 Münster's 1991 Traffic Space Comparison Campaign (Source: City of Münster Press Office)

Willand remarked that a recent trip to Norway showed him the benefits that transitioning mobility away from single-occupancy vehicles can bring. Despite his profession, he acknowledged that the quality of city life improves with fewer cars:

> *I am in the car business. I'm a really passionate car guy. I love to drive. But life is better if parts of cities are car-free.*

SAP's vision for urban mobility scenarios includes the following:

- Car sharing, in which multiple consumers access a single vehicle
- Ride sharing, in which contracted drivers provide rides to individual consumers
- Ride pooling, in which multiple individuals might share a single vehicle in a trip—a more public version of carpooling
- Ride hailing, in which pooling and sharing services are initiated through mobile devices
- Corporate mobility and mobility budgets, in which organizations co-fund their employees' travel
- Subscription mobility, in which individual consumers pay to access a multi-vehicle fleet that could contain, for example, different types and sizes of vehicles
- Micro-mobility, in which smaller vehicles, such as bicycles and scooters, are available to cover short distances

Perhaps most importantly, urban mobility must be seamless and painless. Mixing modes of transportation must become natural and obstacle free: scheduling, switching modes, and paying must all be fluent and integrated to deliver a compelling mobility experience.

Belay emphasized that mobility extends beyond vehicle selection and reaches into the technical infrastructure that enables it:

> *Getting from A to B is a journey, so you better have a map. Back in the day, you needed a printed map. Today, you have a digital map on your phone. Next comes monetization: by zone, by time, by ride, etc. We have to bring all those services into your app. There is also an operational element. If I have a QR code, will the bike unlock for me? If I hop on a bus, will the ticket machine understand me? We have to bring all those services into your app.*

This vision will require a combination of functionalities both for the backend, as described in Table 3.1 and for operations and analytics, as described in Table 3.2, as well as a combination of SAP-led software solutions and partner solutions.

Business Capability	Features
Customer (app)	■ Named and anonymous account management ■ Booking or subscription
Map and route (with partners)	■ Map and information ■ Routing and options

Table 3.1 SAP's Mobility-as-a-Service Vision for the Backend

Business Capability	Features
Utilization and event processing or integration	▪ Identity and authentications ▪ Event capture ▪ Charging right price for use
Face and product management	▪ Products modeling and multimodal product rules ▪ Subscriptions and on-off order management ▪ Own and partner product pricing and charging
Payments and settlement	▪ Payments and credit card management ▪ Partner settlement
Financials management	▪ Credit management ▪ Invoice and billing

Table 3.1 SAP's Mobility-as-a-Service Vision for the Backend (Cont.)

Business Capability	Features
Supply onboard	▪ Operators and routes ▪ Availability and geolocation ▪ Open data (assets and ETA)
Orchestrated routes and products	▪ Binding routes to product bundle or price ▪ Deep linking and embedding with apps ▪ Open API
Apple iOS and Android user experiences (UXs) for customers	▪ Profile and account management ▪ Presentiment and identification ▪ Use and history
Extensibility	▪ Be-in/be-out (BI/BO) services ▪ Vicinity and proximity

Table 3.2 SAP's Mobility-as-a-Service Vision for Operations and Analytics

In the same way that mobile technology has changed navigation and communication patterns among travelers, payment technologies have moved away from tokens individualized to a single transit system or single-use ticket stubs purchased from a platform vending machine. Belay's idea of the payments marketplace includes a variety of vendors:

> *The payments landscape has changed dramatically over the last decade. We used to pay for transit with coins and tokens. Commuters in cities like Hong Kong and London are used to stored value cards like Octopus and Oyster [administered by San Diego-based company Cubic]. The world is moving from cards to accounts, but at a different pace, and with a different technology*

stack. The difference is we, SAP, are coming from the backend. The Cubics of the world are coming from the frontend. We'll meet in the middle.

He sees exciting possibilities for future development of what he calls "be-in/be-out functionality" that would facilitate touchless payments in lieu of high-touch access at restricted doors. "If you have a mobile app, which is native and uses Bluetooth or near field communication, there is no reason why you shouldn't hop on and off the bus without thinking about tickets and such," he said.

Schock compared it to the global telecommunications market:

> *"We're in the network business. How do you reconcile with all the partners? I have a Verizon phone. I travel to Germany, and I'm probably checked in on the Deutsche Telekom or Vodafone network. I don't need to know, it just works. How do they settle all the transactions in the background? The telco guys have figured out how to do this without bothering me, the end user. Well, we are using the same paradigm for our customers in the mobility space."*

SAP's mobility vision to overcome the fragmentation of services and providers is becoming a reality. As Willand described the progress:

> *We had a discussion with BVG in Berlin. BVG is the largest public transport provider in Europe, serving five million people every day. We created a platform called Jelbi.[15] It covers a wide range of mobility options including bus, train, e-moped, e-scooter, bike, car, taxi, and ridesharing.*
>
> *With this platform, you register once. You only give your payment terms once. And you upload your driver's license once. Only one registration and you have access to the whole spectrum of mobility options. Next, working with SAP, we are creating a Mobility Identity, which will allow me to access any form of mobility in the future, not only in Berlin, but also in Amsterdam or Oslo, for example.*

3.6 Choosing Where to Live and How to Move

During the COVID-19 lockdowns, we witnessed empty highways and even emptier office buildings. We saw office workers move away from big cities since they were no longer required to live in high-cost areas to telecommute. Air travel and commuter services suffered steep declines. How will that affect mobility in the long term?

Johnny Clemmons, global vice president and the industry business unit head for engineering, construction, and operations at SAP, noted a temporary reversal in recent years of a broader population migration that has gone on for many decades:

> *Over the last decades, we have seen a huge migration from rural areas to cities. It's really the largest migration in human history, to the point where we're going to have 60% or 70% of the world population living in cities by 2050. It's a significant migration that is changing how people live, work, how they move around, how family works.*

Another consideration is not only where people live, but also how and how often they commute to work. SAP has been observing how COVID-19 has prompted shifts away from office work to remote work, Clemmons continued:

> *If they have individual work to do, where do workers prefer to be? There are people who say they can't get anything done at home because of the distractions. They prefer to go to the office. But for those who prefer to work remotely, it's because it gives them the ability to do the work but still feel like they're a member of their family, and it gives them much more flexibility in the time that they have. You can take 20 minutes and have lunch with your kids or something and then still get the same amount of work done, or even more.*
>
> *It really boils down to flexibility and the commute—those are the two top reasons that I hear from people.*

One consequence is that SAP is re-envisioning its office concept. Clemmons noted some of the structural changes that are supporting employee preferences and organization requirements:

> *SAP has recently redesigned a few of our office spaces post-COVID. We are redesigning them because we're looking at the future offices being mostly used for collaboration, not as much for individual work. We're seeing a lot of companies trying to create this flexible environment where people will spend a couple of days a week in the office, a couple of days a week either at home or on the road.*

Mass transit is gradually creeping back up in large cities, and the move to secondary and tertiary towns is prompting a rethinking of intercity rail in the US. While commuter rail generally has lagged badly in the US, significant progress has been made along these lines recently. For example, Brightline is already constructing its line from Miami to Orlando, and the Texas Central Railway is planning a line between Dallas and Houston. Several other big city connections look promising. The challenge remains: Aviation and good highways provide stiff competition, and raising private funding is always a challenge.

Roland Vorderwülbecke, director of IT at Gebr. Heinemann, a global operator of the Duty Free & Travel Value shops at airports, described how 2020 delivered shocks to the travel industry due to reduced flight schedules, airport closures, and international travel bans. In 2021, some travel resumed, but waves of reinfections bogged down the business. Even a surge in travel in 2022 brought continued challenges, such as staffing problems at airlines, security, airport baggage, and other operations.

Throughout two years of restrictions and surges in travelers frequenting bus stops, train stations, and airports, Vorderwülbecke has observed changes in traveler motivation and purpose:

> *We have also seen a shift in business versus leisure travel. Many companies adjusted to virtual meetings during COVID lockdowns and reduced travel to only essential trips. Individuals, on the other hand, have been eager to travel after two years of lockdowns and so the mix has swung to tourist travelers. However, the significant increase in fuel prices and airfares with the Ukraine crisis could affect this price-sensitive segment.*

He noted that, though domestic travel was responsible for much of the rebound, global travel is likely to recover too, depending on political developments internationally:

> *By 2025, IATA expects international traffic to match 2019 levels, but it will depend on political decisions. If China continues with its Zero-COVID policy, or if Australia and New Zealand again have tight lockdowns, that will significantly impact global travel. Then there is the sustainability challenge for the aviation sector, and electrification is not likely to be a solution anytime soon. For now, airlines are trying in-flight fuel efficiencies.*
>
> *More optimistically, we expect a lot more growth in emerging economies with increasing personal income in populations eager to see the world.*

3.7 The Catena-X Story

Hagen Heubach has observed that one of the biggest problems for automotive OEMs is the lack of visibility beyond their tier-one, or immediate, suppliers. For example, a car company might know the source of its car batteries, but not the source of the lithium mined for the battery. Not knowing much about a supplier's suppliers makes it difficult to anticipate changes in the costs, scheduling, or availability of key resources—and also obscures compliance with company standards or compliance requirements along the supply chain. In our car battery example, a manufacturer with limited visibility into their next-tier suppliers might not know for certain whether the lithium mining company complies with child labor or environmental regulations.

Another industry change afoot is that OEMs no longer sit comfortably atop the supply chain pyramid. Instead, the ecosystem has shifted into a model where more players have equal power.

Like many software vendors, SAP has long maintained cross-industry business networks, such as SAP Ariba for indirect procurement, SAP Concur for travel expenses, and SAP Business Network for Asset Management for equipment and other major capital assets. According to Heubach, the key challenge with these networks is their lack of digital communication between OEMs, first- and second-tier suppliers, and logistics providers because they are unable or unwilling to readily interface their digital systems. The challenges of siloed networks are well understood, but the solution requires a degree of collaboration that had been hard to achieve.

Enter Catena-X, a network specific to the automotive industry designed to create a single uniform data exchange standard along the entire automotive value chain.[16] SAP, BMW, and a number of other major German companies are founders of this partner network. Heubach summarized its goals:

> *With Catena-X, we're getting rid of blockages where the players cannot (or don't want to) talk to each other on the network. We are building an industry business network where SAP can connect to Siemens, SupplyOn, etc., using the same standards. We are opening up rather than trying to be the dominant player, but in exchange we're getting so much more interaction, so much more traction with the market with these value cases.*

SAP is focusing on three key areas:

- Building the technology that enables network communication (in other words, "connecting the dots" between partner networks) using Gaia-X standards for federated and secure data infrastructures.
- Showcasing the use cases that automotive CEOs have been requesting (for good reason), such as end-to-end material traceability, product carbon footprints, or other circular economy details.
- Ensuring fast scalability. Heubach targets having 1,000 partners in the network by the end of 2022.

Currently, use cases for Catena-X are specific to the automotive industry, but as Heubach noted, all these use cases—material traceability through SAP Business Network, product footprint management, digital twins, networked quality management, demand and capacity management, and the SAP Sustainability Control Tower—are equally relevant for other industries:

> *It started with six companies, but we now have more than 120 companies in Catena-X. This initiative goes beyond the classic German companies like BMW*

and Bosch. Japanese suppliers join. Denso is in there. Valeo from Europe is there. Ford has joined. Stellantis is part of the network. We have achieved critical mass.

And it does not stop with the automotive OEMs and their tier-one suppliers. Second-tier suppliers for the automotive industry are giants in other industries: they are chemical companies or mill and mining players.

This is important for my colleagues who look after the business in those industries. If a chemical company discovers and realizes the business value to play in the automotive networks they will ask: 'Hey, why can't we apply the same principles to other value networks?'

Heubach saw potential even beyond manufacturing and is particularly interested in the application of an industry network in the healthcare space:

> *If you squint a bit, a data-driven value chain in healthcare looks a lot like the automotive network. Yes, they have different use cases. They may be less interested in product carbon footprints. There are no recyclers, but there is the patient, and the hospital pharmacy. There are the pharmaceutical companies. There is a ton of special regulation. But the core methodologies, and how you exchange data along a value chain and generate additional use cases and business value—it's really the same.*

3.8 Innovation Ahead

In all the chapters of this book, we describe how industries connect and how classic industry boundaries have blurred. Why should mobility be any different? The "next big thing" in mobility could well be air taxis using electric vertical takeoff and landing vehicles (eVTOLs). A 2022 *60 Minutes* segment[17] described this innovation in the following way:

> *If you've ever had the fantasy of soaring over bumper-to-bumper traffic in a flying vehicle, that may be possible sooner than you think. Not with a flying car, but with a battery-powered aircraft called an eVTOL, a clunky acronym for electric vertical takeoff and landing vehicle. Dozens of companies are spending billions of dollars to make eVTOLs that will operate like air taxis—taking off and landing from what are called vertiports on the tops of buildings, parking garages or helipads in congested cities. EVTOLs promise a faster, safer, and greener mode of transportation—potentially changing the way we work and live.*

Torsten Welte chimed in, as a guest from Chapter 2 on the Everything as a Service megatrend, and added his perspective:

> *Short-distance travel will face a lot of challenges regarding fixed roads. Electric vehicles, however, can go airborne. Great examples from Germany are Volocopter, Lilium, and Wingcopter. It is exciting (for a German, and for mobility) that there is disruptive innovation coming. But of course, there are several in the US, like Joby, Archer, and WiskAero with a major investment of Boeing. This is enough movement to drive innovation to the next level, and I'm looking forward to my first trip.*

However, renowned aviation analyst Richard Aboulafia of AeroDynamic Advisory has said he could buy all the Aston Martins and Rolls-Royces in the world and drive people for free to the airport, and it would be cheaper than developing these eVTOLs and flying them around. But Welte remarked that you can sip champagne in the back seat of a Rolls, but you'd have to be the American president to find a traffic-free highway from Central Park to JFK.

3.9 Integrated Mobility with SAP

Before we close this chapter on integrated mobility, we want to highlight some of the solutions in SAP's portfolio that offer key functionality related to customer centricity, anything-as-a-service business models, connected products, resilient supply chain and smart manufacturing, and sustainability and circular economy.

3.9.1 Customer Centricity

Changing a business model towards sustainable mobility requires several customer interaction points that may be entirely new to the industry, such as direct digital sales without dealer networks and direct customer feedback driving mobility offerings.

- SAP S/4HANA is the backbone for a wide range of mobility product and service providers. It supports processes like order management, vehicle allocation and distribution, accounting of integrated mobility products, service and maintenance management, or profitability analytics.
- SAP Commerce Cloud is the integrated digital frontend platform that handles the transactions between mobility services customers and their providers.
- SAP Digital Vehicle Operations creates a vehicle-centric view for sales and fleet management. This can result in better customer service, better fleet performance, and better cost control.

- SAP Industry Process Framework enables efficient application integration to create processes that deliver mobility services while requiring access to multiple SAP and non-SAP systems.
- SAP Billing and Revenue Innovation Management enables scalable billing of sophisticated pricing structures based on product and service bundles, subscription models, or usage-based agreements with metered consumption data.
- SAP Configure Price Quote enables sales executives and their support teams to tailor anything-as-a-service contracts to their customer needs, ensuring that products and services can be profitably delivered and priced.

3.9.2 Anything as a Service

Integrated mobility combines products and services to deliver mobility services tailored to the preferences and needs of their target customers. The data and process platforms of the involved providers need to be integrated to sell and deliver mobility services with a seamless customer experience. This transformation is also a cultural change for the established mobility product and service providers, driven by a customer-centric mobility design. Integrated mobility platforms combine competing modes of mobility: train rides compete with air travel, e-scooters compete with taxis, ride-hailing competes with rental cars, parking at airports competes with shuttle services.

The following SAP solutions enable mobility product or service providers to deliver their services efficiently and conveniently, directly or through service consolidators on mobility platforms:

- SAP Billing and Revenue Innovation Management enables the definition and operation of complex pricing structures to settle and bill product and service elements as part of a comprehensive mobility service offering.
- SAP Configure Price Quote enables sales executives and their support teams to tailor mobility product and service contracts to their corporate customers, while ensuring that products and services can be delivered at competitive yet profitable price points.
- SAP Commerce Cloud and SAP Sales Cloud enable contract management with their service level agreements and interface with customers and service delivery entities.
- SAP S/4HANA processes the fulfillment functions of mobility service contracts, including materials management, financial accounting, and maintenance.
- SAP SuccessFactors manages the people side of delivering mobility as a service, including recruiting, training, and payroll functions.
- SAP Fieldglass makes mobility service organizations scalable by extending the workforce dynamically with temporary labor.

3.9.3 Connected Products

Integrated mobility requires reliable and compliant vehicles in the right location and in the right condition to deliver the envisioned user experience. The fully integrated digital twin of a vehicle mirrors its production, operations, and maintenance history. These digital twins can be integrated into a complex, digital, real-time representation of the mobility system, including the users on the move.

The following solutions enable innovation around connected products:

- SAP Industry Cloud Digital Vehicle Hub models and manages the digital twin of vehicles and their key components.
- SAP Business Technology Platform is the extension and integration platform for connectivity data that build the elements for operating a connected mobility system in real time.
- SAP Open Vehicle Connection Hub integrates telematics data into business processes like predictive maintenance, billing, insurance, recharging, or refueling.

3.9.4 Resilient Supply Chain and Smart Manufacturing

Integrated mobility relies on a range of vehicles from connected e-scooters and bicycles, privately owned and fleet cars, trams, trains, and planes. Sourcing components, managing complex supply chains, and assembling vehicles requires top-floor to shop-floor integration, advanced production logistic processes, and digital intercompany processes along the entire automotive value chain.

A broad range of SAP solutions contributes to resilient supply chains and smart manufacturing:

- SAP S/4HANA is the backbone for best-in-class automotive logistics processes like just in time, just in sequence, Kanban, and advanced shipping and receiving.
- SAP Returnable Packaging Management supports a circular flow between suppliers and OEMs of returnable and reusable packaging materials like containers or pallets for improved sustainability and cost reduction.
- SAP Quality Issue Resolution enables collaborative defect and resolution handling in the automotive industry using the 8D methodology.
- SAP Business Network provides functionality for material traceability in multi-tier supply networks.
- SAP Collaborative Demand and Capacity Management balances demand and supply in the network of OEMs and their suppliers.
- SAP Digital Manufacturing Cloud analyzes global and plant-level manufacturing performance and creates more process visibility to maintain productivity and quality levels.
- SAP S/4HANA Supply Chain for extended service parts planning optimizes the spare parts acquisition through effective planning and distribution.

3.9.5 Sustainability and Circular Economy

Financial investors, customers, and regulators scrutinize mobility providers for the sustainability of their products and services. SAP offers a broad spectrum of solutions to live up to sustainability-related regulations and to apply best practices for maximizing circularity, reducing greenhouse gas emissions, minimizing environmental product service footprints, and implementing extended producer responsibility initiatives for the entire supply chain.

- SAP S/4HANA Environment, Health, and Safety Management generates enterprise-wide real-time awareness for non-compliance and supports a modern safety culture for employees and the environment.
- SAP Product Footprint Management analyzes the environmental impact of products and services along their lifecycle for standards-based disclosure, and drives the effort to reduce greenhouse gas emissions and other environmental factors.
- GreenToken by SAP enables the trusted exchange of carbon footprint data along the value chain, from raw materials and energy through the usage and disposal phases.
- SAP Responsible Design and Production uses live connectivity with SAP S/4HANA data sources to enable products that can be easily repaired, recycled, and re-processed in a circular economy.
- SAP Analytics Cloud creates holistic insights and reporting across all operations. This analytics platform combines data elements from internal and external sources for better insights and more informed decision making.

Chapter 4
New Customer Pathways

From the Bazaar to Amazon

If you visit a souk in Cairo, a bustling marketplace in Hong Kong, the fish market in Tokyo, or just a weekend farmer's market near you, you'll feel the pulse of commerce as it has been beating for millennia. World history and global logistics have long been influenced by trade routes for silk, spices, amber, and other valued commodities. Marco Polo traversed the Silk Road all the way from Italy to China. Our modern-day corporations have been influenced by the global trade patterns of British East India Company and the Dutch equivalent, Vereenigde Oost-Indische Compagnie.

Apple today has suppliers around the world, shipping components to Foxconn plants in China—and in reverse, iPhones, iPads, and other devices are then distributed to consumers who buy them online and in stores around the world. Yet, the fastest-growing component of Apple revenues comes from services and apps, which are mostly sold online and delivered digitally. Similarly, BMW assembles all its SUVs in South Carolina and ships them around the world, making the city of Spartanburg a major multimodal logistics hub. While most of BMW's sales are still channeled through dealers, it increasingly sells subscriptions online for many of its auto features. Amazon is known as the largest e-commerce vendor in the world, and owns a massive network of distribution centers, delivery operations, and reverse logistics infrastructure. It has increasingly been opening brick-and-mortar stores, such as Whole Foods, Amazon Fresh, and Amazon Go.

Nineteenth-century Philadelphia retailer John Wanamaker allegedly complained, "Half the money I spend on advertising is wasted; the trouble is I don't know which half." These days we brag that we can effectively measure payback from digital campaigns, but that's not the whole picture: A third of the world's advertising money is still spent on billboard, direct mail, print, and other analog media.

SAP has over 13,000 retail and countless consumer product customers in over 100 countries. None of them are exactly alike. They are carving a dizzying range of pathways to their customers. In this planet-scale laboratory of experimenting with new customer pathways, we'll search for patterns and discuss the outlook for retailers and their technology providers like SAP.

4.1 Blurry Industry Boundaries, Crisp Customer Segmentation

Let's take a look at the overall consumer industry space, starting with a view of how companies and industries are responding to shifting customer expectations and demands and how consumers play their part in shaping the trends.

Matt Laukaitis is executive vice president and global general manager of SAP Consumer Industries. He offered SAP's definition of consumer industries:

> *Consumer industries, in our former definition, included retail, consumer packaged goods, and wholesale distribution. A few years back, we also added healthcare and life sciences. Then earlier this year, we expanded our scope to agribusiness because we see so much opportunity across channels as companies operate beyond traditional industry boundaries. In each of those sectors, we're seeing significant conversation about building pathways directly to the consumer.*

SAP observes blurring industry lines and blending business models. Broadening our view on consumer industries helps us discover the relevant trends that shape our customers' strategies and their need for innovative digital solutions. We try to give each of our customers the opportunity to optimize their business and take advantage of the growth opportunities that might exist in an adjacent or even unrelated industry. Serving the full range of consumer-oriented industries with solutions specific to their core business allows SAP to spot patterns evolving from experiments into the next industry practices.

Achim Schneider is global head of SAP's retail industry business unit. He provided more examples of blurring of industry boundaries, citing one organization's expansion to an adjacent industry as a customer experience mechanism and organic branding opportunity:

> *IKEA operates their own restaurants. Initially, the basic cafeteria provided a convenience for exhausted shoppers with their cranky children. The cafeterias with their Swedish menu and the iconic Köttbullars meatballs are an integral part of IKEA's brand, reminding patrons of IKEA's Swedish heritage in every store around the world.*

Meanwhile, he said, other businesses are capitalizing on a diversity of categories and offerings, making such cross-business synergies integral to their business models. Retail, healthcare, dining, and other types of businesses are blending together and dissolving traditional distinctions. Schneider shone a light on several examples of cross-fertilizing businesses:

> *Sansibar started as a small company on an even smaller island, Sylt, in North Germany. They run fashion stores and have their own brands. They sell over 1,500 wines at stores across Germany. They have a hospitality and catering business, and they run fancy restaurants. I think that kind of combination is becoming more and more important for these hospitality retailers. On the other side of the planet, Jumbo Seafood in Singapore is moving in a similar direction. And a huge Korean restaurant chain told us 'Well, we think we are more of a retailer.'*

Laukaitis described some of the high-level trends he has seen in the last couple of years and what fuels those trends:

> *Customer expectations are continuously rising. As consumers, when we go with another brand, we expect the previous experience plus so much more. In the continuous struggle to attract and retain consumers, brands have to deliver more and better-differentiated experiences. What started with a coarse market segmentation to identify and serve different consumer groups has become a data science that narrows segmentation down to a single person. Now you just need to figure out how to run such a business profitably and at scale.*

Laukaitis pointed out that the COVID-19 pandemic has changed the attitude towards innovation and risk:

> *Companies got addicted to rapid innovation. In the past, there were a million reasons why corporate inertia would say, 'You know, we've thought about a curbside pickup, but we don't need to do that now. We've got so many other priorities.' But many of our customers surprised themselves when they were pressed to change: they could get new services like curbside pickup done in a couple of weeks. Now the leadership teams of companies and their boards have come to like the new speed, saying, 'Okay, now we know we can do things very quickly, so let's start knocking down this list that previously looked kind of unassailable.'*

Importantly, brands must make strides in the midst of supply chain breakdowns, massive labor turmoil, and shifting customer expectations:

> *In the past, the world was all about driving the right promotion planning and about aligning and balancing the supply chain. While this remains critically important, we have all seen that there is a significant challenge to actually*

responding to supply chain disruption. We provide capabilities to our customers so they can be very, very flexible in their understanding of multiple levels of the supply chain to help shape customer expectation to what they can actually deliver.

The Rolling Stones sang it nicely: "You can't always get what you want." Industry leaders are writing the next line: "So we make you want what you can get."

The consumer industries are not alone with the huge talent disruption dubbed the "Great Resignation" in the United States. The financial services industries, the public sector, and energy companies share the pain, but lower-skill jobs are particularly at risk. Growing customer expectations, disrupted supply chains, and an eroding workforce make SAP customers approach us and our partners to define strategies about how to prioritize their investments and how to use technologies to soften or avoid blows to their businesses.

Accenture's design agency, Fjord, coined the term "liquid expectations" for the idea that customer experiences are fluid across industries. Laukaitis agreed with this notion:

> " *Consumers don't care if a brand is owned by a traditional consumer products company, by a retailer, or by a healthcare or financial services company. They go for products and services that meet their needs and preferences, and don't care about historic industry boundaries. This also motivates players to venture into new territories and to experiment with disruptive business models.*

In recent years, consumers have also become more aware of the societal and ecological impacts of their buying decisions and consumption habits. They prefer to buy from companies that conduct business in a way that's respectful of the environment and the limits of our planet. This trend compels consumer product companies and retailers to demonstrate their trustworthiness by showing their efforts towards circularity, curbing emissions, and fair working conditions along their supply chain. Commitments to "net zero" and the need to comply with regulations related to extended producer responsibility (EPR) complement the need to drive product and process innovation in response to changing demand patterns and consumer preferences.

Thinking with the Circular Economy megatrend in mind is especially critical as the number of returned products has grown significantly—particularly for products ordered online. Focusing specifically on the retail space, Schneider outlined four key trends:

- Commerce is everywhere, and omnichannel retail is a lot more than a buzzword. "From in-store digital displays to online marketplaces and popup stores, today's omnichannel retail eliminates the delineation between channels. To compete and thrive, retailers must provide seamless shopping experiences," he said.

- Customer convenience is fueled by the "everything from home" trend during COVID-19. It has become table stakes for retailers, Schneider said. "Curbside pickup, 'buy online, pick up in store,' 'scan and go,' and 'just walk out' options are here to stay, allowing customers to shop and collect goods conveniently. Micro-fulfillment centers and partnerships with last-mile companies, such as Uber and Lyft, offer additional customer-pleasing and cost-effective options—all made possible through smart technologies."
- Micro-segmentation and personalization to individual consumers are increasingly affordable—and necessary, Schneider added. "Powerful retail analytics can shed light on customers' current and future needs, helping retailers shape their offerings. Personalized marketing and experiences—based on consent and trust—build customer loyalty and, ultimately, profitability."

  By engaging intimately with individuals and correlating data from multiple sources, including social media, organizations can learn a lot about their customers' preferences and habits. Adrian Nash, head of strategy for the SAP Customer Experience product portfolio, commented on personalized data and consumer consent:

  > *From a data privacy perspective, I think we're one of the largest stores of consent, so opt-ins, opt-outs, renewals, revocations on the planet with around 13 billion opt-ins, opt-outs across customers like sporting associations to consumer products companies. You can say 'I want to get this email' or 'My favorite team is so-and-so,' and then it'll start to personalize communications based on your first-party preferences and consent rather than from third-party knowledge and hearsay.*

- Sustainability has become a driving force. Schneider added to the comments Laukaitis made about the circular economy and purpose-driven retail trends, emphasizing the positive impacts to be derived by customers and businesses that practice corporate social responsibility (CSR):[1]

  > *The retail industry is under scrutiny from both consumers and regulators, as retail companies are a major contributor to carbon emissions. At the same time, retailers are in a powerful position to enforce sustainability principles across their supply base. Leaders are thinking about environmental and social governance standards not as problems but as opportunities to achieve more efficient and less wasteful practices that safeguard their license to operate.*
  >
  > *The next generation of customers is looking for brands that align with their values, and they are willing to 'walk the walk.' Commitment to a purpose has become as important as a brand's digital experience.*

4.2 The New Map to Reach Customers

Ralf Kern is global vice president of SAP's retail business unit. He described how customer expectations vary with the retailer they engage with:

> *If you shop in a convenience store, you're happy if you're in and out within a minute and you get exactly what you need. At a fashion retailer, like Louis Vuitton, you may have a glass of champagne, and stay a couple of hours and expect service to get your selection right. At a DIY store, you may want to rent an appliance and get it delivered to your home and have a painless return process, and if you find that you couldn't make the bathroom tiles stick as well as you thought you would, they can give you advice or the phone number of a reliable craftsman. In Germany, if you go to an Edeka, you expect a broad selection of fresh grocery compared to an Aldi, a Lidl, or another discounter.*

Kern also offered his perspective on personalization and omnichannel experience, emphasizing the need to be data based and to gather such data in a trusted way:

> *Today, we talk about hyper-personalization, fantastic customer experience, and so on. There is a lot of omnichannel, omni-commerce talk. It all comes down to the fact that you can personalize promotions and your customer communication, and how you curate your assortment to keep your customers curious. But you can only personalize if you have the consent of the customer—legally. You have to do it in a trusted way.*

Laukaitis spoke about another important wrinkle in the retail fabric: franchise stores, which have distributed ownership that can result in customer experience variations:

> *Żabka Polska is a chain of convenience stores in Poland. Similar to Wawa in the United States, customers of Żabka come in multiple times a day and expect a great consumer experience every time.*

Żabka has a highly franchised and distributed model. To address the orchestration of huge data volumes, Żabka selected SAP Business Technology Platform as the integration backbone. This allowed Żabka to implement a centralized franchisee relationship management system and cloud e-signature solution with SAP's integration orchestrating the communication between the participating systems.

> *Through our partnership, Żabka achieved a seamless franchisee experience of the entire chain and unified data at headquarters, which resulted in a 15% better franchisee experience and 20% more efficiency in franchisee recruitment.*

I think Żabka is a really interesting example of taking a holistic view of the business, regardless of how it's structured from a corporate perspective.

Figure 4.1 shows the wide range of pathways to companies beyond going direct-to-consumer (D2C); the many different forms of business-to-business transport, the key roles of wholesalers/distributors and retailers and the many sales options consumers can choose from. Successful consumer product companies manage to hide the complexity of the system from the consumers by delivering a unified and engaging experience. Seamless channel orchestration for sales and fulfillment is only one part of the customer journey: Intelligent, fair, and convenient returns and reimbursement processes, warranty and repair management, recycling, and refurbishing...all these processes are even more complex and have a lasting effect on customer experience and loyalty.

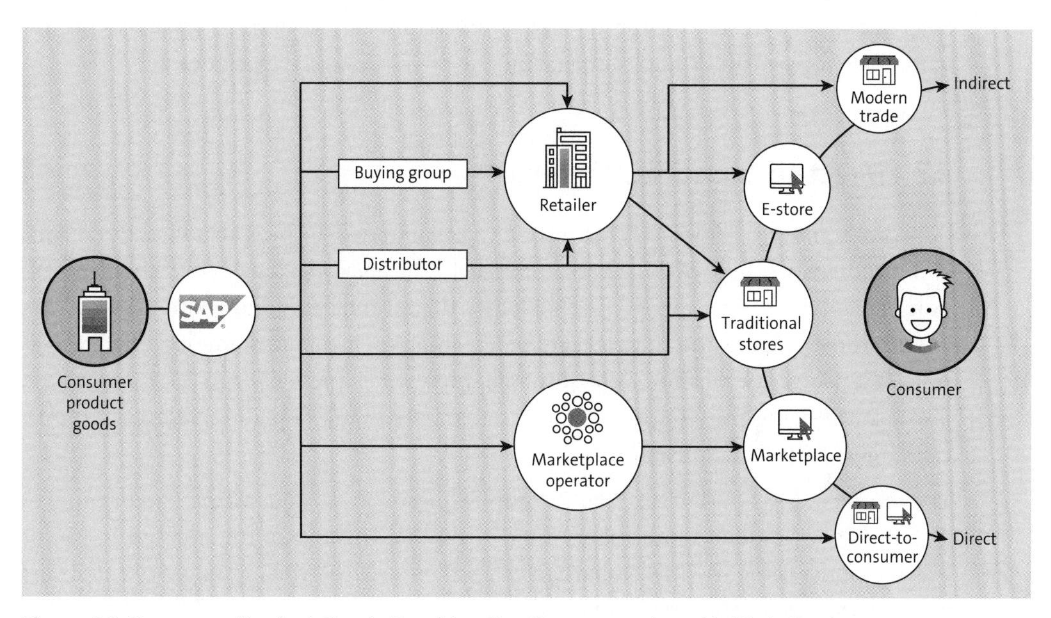

Figure 4.1 Consumer Product Goods Reaching the Consumer along Multiple Routes (Source: SAP)

Nash noted that companies need to pick and choose from a wide menu of solutions to support the network of vending and servicing pathways:

> *We have a portfolio of products that serves many use cases: direct to consumer and all the flavors of indirect sales and fulfillment. Actually, depending on which business you're in, there's a certain cluster of the solutions that make sense for you. In a grocery store, you probably don't want a salesforce automation tool but you may need a customer engagement engine with built-in personalization.*

For Nash, the network between a consumer product company and their customers may seem nice and orderly, but the individual routes to market easily get isolated, making cooperation pathways almost impossible. He saw a challenge in that many companies end up with a siloed view, implementing processes and systems with the objective to cover the needs of each route to market without looking for synergies. With this approach, a company could end up with dozens or even hundreds of applications sending data all up and down the route but without effective crosslinks: Too much inventory in one channel may collect dust or go to waste, while another channel disappoints the consumer with an empty shelf. Large consumer products companies can have an even bigger problem as collaboration between brands or subcompanies is hard to manage if the technology landscape grows by acquiring more brands or opening new routes to market.

In the upcoming sections, we'll focus on some popular pathways companies have pursued as they try to stay on top of their omnichannel strategy: D2C avenues, the "re-commerce" opportunity, and reimagining the physical store experience.

4.3 Direct-to-Consumer Outreach

Consumer product companies have tried to sell and deliver products directly to consumers for ages—think of the iconic milkman or the *Encyclopedia Britannica* salesman. Companies have long been opening their own digital and brick-and-mortar stores. Taking out the retailer as the middleman and keeping their margins are obvious motivations; a less-discussed factor is consumer privacy, which has led more US states to pass laws against using third-party data and to mandate that companies collect data from consumers with their consent.

The recent success of startups like Peloton for home gyms and Warby Parker for eyeglasses has accentuated interest in the D2C model. The appeal for startups to target every product category (like Dollar Shave Club did with a growing range of toiletries) was well captured in *Inc.* magazine:

> “ ***By selling directly to consumers online, you can avoid exorbitant retail markups and therefore afford to offer some combination of better design, quality, service, and lower prices because you've cut out the middleman. By connecting directly with consumers online, you can also better control your messages to them and, in turn, gather data about their purchase behavior, thereby enabling you to build a smarter product engine. If you do this while developing an 'authentic' brand—one that stands for something more than selling stuff—you can effectively steal the future out from under giant legacy corporations.***[2]

Schneider pointed out, however, that D2C is hardly a new trend and that additional expenses for shipping and handling, for example, might make some products more profitable for D2C channels than others:

> *Adidas, Nike, Under Armour, and others have been going direct for a long time. Adidas has grown to 12,000 stores since it entered China in the late '90s. I think that was a very clear indicator how many consumer packaged goods companies are thinking. If I have a Nike store, it could be a franchise store or it could be an owned Nike store. For the consumer, it's the Nike brand. However, it might be a bit different in cosmetics or personal care brands going direct to customers.*

Nash shared an example how innovative D2C routes for coffee capsules run parallel to classic routes for chocolate powder:

> *Some of the CPGs we work with have more than 2,000 brands; some are leading the way in direct-to-consumer channels, but it really depends on the cost and service of the solutions they provide. You may not want to subscribe to a hot chocolate brand online, but you may want regular coffee pods of your favorite flavor delivered to your door.*

He observed some consumer packaged goods (CPG) companies pick a brand that has potential for a subscription model, like Gillette razor blades, and use that as a launching pad for a direct connection with consumers. That's one way to start populating SAP Customer Data Platform, which Nash called "today's CRM book of record":

> *It not only knows who the customer is with the attributes that you would expect to know about a customer. It also knows that they have opened an email, launched an app, added something to the basket, or walked into a store. It's all of the unstructured signals that can better influence how you engage with that customer later on. That's the value of our customer data platform. It works across engagement points to trace and predict the customer journey, ready for the next engagement.*

Laukaitis shared the example of how Casey's General Stores enriched its customer database. The company has a presence in over 2,400 locations in 16 US states.[3] Organized as self-service gas stations attached to convenience stores, Casey's offers grocery items and prepared foods such as pizza and sandwiches.[4]

Using the SAP Customer Experience portfolio, Casey's streamlined its online ordering platform, introduced a mobile app, and created a new loyalty program. Laukaitis commented:

> *After experiencing increased demand for its pizza and grocery items due to COVID, Casey's partnered with SAP to streamline their online ordering platform in a matter of weeks. They also updated their mobile app and loyalty program so they can truly focus on guests, better understand their needs, and deliver experiences that exceed those needs. The app was wildly successful, with three million downloads within six months, and Casey's very quickly generated over 65% of its digital revenue through the app. While Casey's had an understanding of who was actually buying gas and shopping in the store pre-pandemic and pre-app, they now have enriched customer data and patterns.*

Laukaitis pointed to more D2C opportunities as companies move across industry lines:

> *One of our retailers bought a home dialysis company, so we're helping them integrate that business efficiently. We have big market share in the life sciences industry, so we're sharing our experience with how the life sciences industry interacts with the clinical world in delivering drugs, and following up on the patient's experience with the drug. So, it's an exciting area for us because we can add value for our customers with our broad industry expertise.*

One D2C challenge is how to compete with the scale of fulfillment logistics of an Amazon or a Shopify. Many consumer companies are used to shipping pallets, not single items. Going with specialist third-party logistics companies for order fulfillment has been the solution for some. As we discuss in Chapter 9 on the Resilient Supply Networks megatrend, support for "micro-fulfillment" has emerged, including smaller warehouses closer to consumers, warehouse bots that can be deployed as a service, and last-mile delivery ecosystems. All these approaches can reduce the historical fulfillment barrier to online D2C commerce.

4.4 The "Re-Commerce" Opportunity

Returns are an organic part of commerce but are still painful for many companies because they represent a reversal of revenue-generating sales and incur significant additional process costs. Returned goods often involve a customer service inquiry and must be refunded based on predetermined, curated criteria. Return shipping and delivery back to the warehouse require time and resources. Finally, the items must be reshelved, refurbished, or disposed of. Each of these process steps hurts profitability.

The increase in online sales has also contributed to an increase in returns rate. In 2020, the return rate was at 10.6% but soared to 16.6% in 2021 when US retail sales reached $4.583 trillion.[5]

While returns have always been a business reality for retail organizations, the uptick in the returns rate is worrisome. Laukaitis commented on how organizations can respond, citing one online apparel and footwear retailer who he believes applies best practices for returns management:

> *Zappos and other retailers have conditioned consumers to expect a hassle-free returns experience. It is important to consider both the employee and customer experience in this process and understand how to optimize margins for the business. It's essential to empower the people at the customer interface to handle items and guide the consumer through the returns process. Then it's important to educate the store and warehouse employees on how to inspect the returned products and to decide on the best routing of the returned product.*

As in many areas, Amazon has blazed a trail by making returns "almost too easy" for customers. Even its website offers the simplest of instructions:

> *Don't worry about printing a label or packing up your item. Just go to Your Orders and select the item you wish to return. Tell us why you want to return this item, then choose a drop-off location that supports label-free, box-free returns. We'll send you a QR code; bring your code and item to the drop-off location, and you're done.*[6]

To determine the best fate for returned items in the "reverse logistics" process, third-party logistics players apply artificial intelligence (AI) to support or even make such decisions. goTRG is one of these providers who process items on behalf of retailers and determine whether a returned item is fit to go back on the shelf, can be refurbished, or must be scrapped. Centralized returns management benefits from the economies of scale, although stores have tried with little success to deploy standalone technologies and even manual systems.[7]

The customer experience can be miserable when buyers try to return items at the store. As Kern observed, switching from selling to returns processing is a disruption for store assistants whose frazzled nerves may lead to poor customer outcomes. Not only are they spending "unproductive" time on handling the return; they also have to make difficult decisions about returns, such as whether to accept the return and what happens to the returned item. Best practices also vary with the type of product. Laukaitis used SAP customer Tapestry, which has luxury brands such as Coach, Kate Spade New York, and Stuart Weitzman under its umbrella, as an example:

> *Luxury brands have returns policies that are pretty generous because they want to be customer-focused. But how do they make sure every return makes sense? The speed of adjudicating the return is critical. The verification process obviously has to be pretty tight, given the high dollar value of those products. For example, are there particular products within the Coach portfolio that cause more trouble from a return perspective, or are there certain things that the associate has to be trained to look for before they actually accept it at full value? Provenance and authenticity are vitally important to the brand, so it's essential to avoid restocking counterfeit items.*

SAP recommends SAP Intelligent Returns Management to make the best out of the customer interaction. Laukaitis suggested accepting returns as a natural element of the retail business but also shifting perspectives and trying to turn returns from unhappy customers into new sales opportunities:

> *We are having conversations with great retailers about their luxury and high-volume brands. Many of these brands want to take back control of the post-purchase experience. They feel they are missing an opportunity if they're totally disconnected from the subsequent purchase by somebody else.*
>
> *The luxury brands try to keep the upper hand in the struggle with secondary markets like Rent the Runway that allow you to rent designer clothes. Some platforms have disintermediated their brands because they want to provide their customer with a dedicated full lifecycle experience. If returns handling, refurbishing, or restocking are not done in the right way and consistently with the brand strategy, that's a problem for them. We're trying to help them take control of that conversation.*

Schneider commented that large return volumes are forcing companies to think about reducing the number of returns rather than just optimizing the returns process:

> *How do we make our customers happier, provide them the best experience, and make it convenient for them? Convenience means, 'Oh, I'm happy with my purchase. It fits. The color is nice. Perfect shopping experience.' Convenience also means, 'Oh, I don't need to go back to DHL and return the product.'*

He suggested we should extend our focus from just optimizing returns processes to finding innovative ways to avoid returns, which makes the customer happier and the business more profitable:

> *There are some 20 key questions a retailer can ask themselves to reduce returns. Here's an easy one: Is the online product information about sizes, fit,*

> *and colors accurate? Can I add logic to help consumers find the best product for them? What if a 3D-scanned digital twin of the customer can try the item on in a digital fitting room, so the analog twin can make a better-informed buying decision?*

Better product descriptions would definitely help. Specright is a startup in the SAP.io Foundries ecosystem. In an interview, CEO Matthew Wright said it is time to move to a "specification-first world."[8]

> *We have seen an explosion in SKUs and growing complexity of supply chains. Today's industrial technology was designed before complexity grew exponentially, and the lack of high-quality product specifications is the root cause of many supply chain issues.*

We can only agree with Wright, and a look at the sheer volume of returned goods, both from a value but also from a volume perspective, shows quite clearly that each tenth of a percentage point reduction in returns can be great news for the bottom line and for the environment. Better product specifications can make a significant contribution in this regard.

4.5 Improving the In-Person Experience

In an analysis from March 2022, McKinsey identified "five zeroes" that are transforming the role of stores:[9]

- Zero difference in channels
- Zero desire for assistance with transactions
- Zero wait times for delivery
- Zero tolerance for inaction on equity and sustainability
- Zero wiggle room on talent

Let's look at the call for zero wait times in delivery. McKinsey observed that "one in three shoppers now expects same-day delivery—a share that will grow as powerful retail giants and nimble startups compete in the realm of customer experience." Zero wiggle room on talent means that highly trained and digitally supported personnel is required along the full value chain:

> *A great customer experience, in store and online, now requires end-to-end product availability and visibility, which in turn requires first-rate operations: talent in IT, logistics, procurement, fulfillment and other functions. Similarly, in-store staff need to recognize when customers need help—and when they*

don't. And with high employee turnover and the need to keep up with rapid changes in the marketplace, retailers must train new hires and longtime employees quickly, effectively, and at scale.

This pursuit of high-quality, intuitive, and customer-centric store experiences has naturally changed how stores are organized. Schneider provided us an idea of how retail stores have been morphing:

> *Stores are becoming small logistics centers. It does not matter if they are grocers or hard goods retailers. Customers are ordering online and picking up in-store. How can you ensure that you have the right product at the right time in the store if you promise a less-than-two-hours pickup time? In fashion, stores are becoming experience centers. You get the consultation and you can see, touch, and feel the fabric. You order, and you get it shipped home. Stores don't have to have all the colors and sizes of the products available. You just have the selection there.*

Some stores are using technology to collect consumer data and facilitate purchases using AI from data pulled from multiple sensors. As Schneider illustrated with some examples, stores are changing their operating hours in response to a varied purchasing experience:

> *There is a Swiss retail chain called Valora with a small footprint near train stations. They are creating the convenience store of the future. Customers can shop around the clock and at their own pace. Their smartphone unlocks the store and pays the bill. When shoppers take a product off the shelf, after a short scan it appears in the digital shopping basket. Once customers leave, they are billed at the touch of a button. The number of store assistants engaging with shoppers: zero.*
>
> *Livello, in Germany, is another example. They pilot many innovations in their stores to see, 'Okay, what can we do to deliver a better service to our customer? The Livello store is especially popular on Sundays, during holidays, and after hours as it is the only available supply option that is always open.*[10]

Given severe labor shortages in so many industries, we don't hear arguments about replacing human labor with technology.

Another set of store-based technologies that is changing the way consumers purchase are augmented reality (AR) and virtual reality (VR). For example, these technologies allow consumers to virtually walk through a store or to even digitally "try on" a garment. On its website, Nexttech AR mentioned the acceleration of AR adoption during COVID-19:

> *Although AR shopping has been around for a few years now, many retailers remained skeptical—until the closure of all nonessential physical stores, that is. In a matter of days, the in-store shopping experience ceased to exist, forcing many brands to seek new options to help their customers find a better product fit while shopping online.*

Given the merge between AR and VR, a new term—*mixed reality*—has arisen. Schneider talked about other store-based technologies, such as smart mirrors, and pointed to the retailers Burberry, Under Armour, and Adidas as organizations piloting innovations in this space.

SAP has a focus on smart stores at the SAP Experience Center at Walldorf, where it showcases some of the technologies you can experience:

- Devices for counting people to analyze traffic into and out of the store and monitor key passageways like stairs and elevators
- Demographic cameras pointed at an entrance to register the age and gender of consumers
- Multi-3D sensors that detect dwell time and route
- Cameras directed at shelves to detect when a product is no longer available—perhaps selected, replaced by customers in the wrong spot, or out of stock
- Electronic shelf labels that can be updated with new prices in real time

Schneider cautioned that "technology is not automatically always the magical solution for everything. It's important to consider where and how it makes sense." He pointed to the balancing act companies must maintain as they evaluate store technology. He shared a story of an executive who visited the SAP Experience Center at Walldorf and commented, only half-jokingly, "Do you know how long these cameras would last in our stores? Exactly one minute. The second customer who comes into my store will walk out with that camera." Loss prevention applies both to products and to in-store technologies and must be integrated into store concepts.

Many SAP customers are already sharpening the consumer experience in different ways. Ritu Bhargava, president and chief product officer of SAP Customer Experience, described her recent visit to a Coop store in Switzerland. For Bhargava, the Coop store experience was streamlined and enjoyable:

> *Imagine yourself waking up in the morning and having to go grocery shopping. You're probably dreading it, thinking 'Oh, dear, I have to go grocery shopping.' Well, even before you're thinking about the visit, Coop has made groceries fun and relevant. They have been sending you offers and promos. They have a companion loyalty app, personalized to you. They know you buy hair or shaving products, so the promo codes are personalized to what you buy.*

Where once customers might have been inconvenienced by paper coupons printed on receipts, they can now access customized discounts on their mobile devices whenever they're ready for the next purchase.

In addition, Bhargava said, stores cater to customers personally and meld the physical with the digital:

> *They've also blurred the lines between physical and digital shopping. From the moment you drive into the parking lot, they recognize you from your loyalty app, based on consent you provided. The experience of how you get to park, how many hours of parking you get for free, whether you get preferred parking, starts from that moment. When you physically enter the store, your companion app also processes your purchases and checkout, which means that you can use your phone as a scanner. As you start scanning your grocery items, the promos that you were sent get applied in real time. That is an engaged front-end experience.*

Customers also find that Coop provides a true omnichannel experience and an incentive program, said Bhargava:

> *Integrating the physical and digital world means they can power different purchase and return options, like buy online and return in store. Of course, they have loyalty points. But they're taking it beyond that to make it experiential loyalty, which is no longer about just earning and burning points. Then when you step out of the store, there is a freebie, like a stuffed toy or a free gadget, waiting for you to pick up. You'd be surprised how many kids and adults want to go to the store because they are going to get that little toy at the end of their checkout experience. There are also wine clubs with in-person events. There are outdoor activities clubs that promote healthy living. They're going after delightful experience at every moment, and scaling it to a complete experience.*

We believe that we're still in the early stages of a profound transformation in how consumer industries view their customers. Technology has matured to a point where creating immersive digital experiences is ready to move from the lab into the wild.

4.6 From Managing Relationships to Shaping Customer Experiences

The concept of customer relationship management (CRM) has been around for three decades. For a long time, technology has not really improved customer relationships that—as the acronym suggests—apparently need managing. Shifting to the "customer experience" (CX) moves the focus from doing something to the customer into creating an environment that evokes customer experiences. But a single acronym still can only be a high-level label for a range of capabilities SAP customers require to create the settings for a delightful customer journey along innovative pathways. Nash clarified his CX perspective:

> *I interpret CX—as opposed to core CRM—as a mix of two domains. It's the digital experience domain (which is usually content, commerce, identity management, profile management, and some marketing and service) to the CRM, which is typically marketing, sales, and service. It's an amalgamate of these two with a focus on the customer perspective.*

In his strategy role, Nash exchanges views and trends with industry analysts and monitors the competitive landscape. In conversation with customers, he talks about five major solution categories: customer data, commerce. marketing, sales, and service—and demonstrates how they apply to different touchpoints in business-to-business (B2B) and business-to-consumer (B2C) settings, as shown in Table 4.1 and Table 4.2. Nash labeled one column "Broader SAP," which in many ways is even more critical than the other columns because it represents a broad foundation of capabilities, from product design and manufacturing, to sourcing and procurement, human resources, finance, and fulfillment—everything required to enable the experience SAP customers want to evoke.

Category	Customer Data	Marketing	Commerce	Sales	Service	Broader SAP
Thought leadership videos		✓				
Webinar	✓	✓				
Contact form	✓	✓		✓		✓
Sales engagement	✓	✓		✓		✓
Quote and negotiation	✓	✓	✓			✓
Product received	✓	✓				✓
Loyalty and self-service sign-up	✓	✓	✓			✓
First service engagement	✓	✓			✓	✓

Table 4.1 B2B Engagement-Driven Processes

Category	Customer Data	Marketing	Commerce	Sales	Service	Broader SAP
Social discovery	✓	✓				
Social shop			✓			
First purchase	✓		✓			
E-receipt loyalty signup	✓	✓				
Delivery update	✓	✓				✓
Product delivered	✓	✓				✓
Product offer recommendation	✓	✓				✓
Self-service support	✓	✓			✓	✓

Table 4.2 B2C Engagement-Driven Processes

Laukaitis shared a number of examples from around the world to expand on the "Broader SAP" category:

- **Salling Group, the largest retailer in Denmark**
  "We have a number of different innovation initiatives with them. One key initiative was around the amount of time their store managers spend running the store on a day-to-day basis. We helped the customer, who has over 50,000 employees, optimize the administrative process around open positions from a recruiting perspective through a digital transformation of their mobile recruitment and talent processes."
- **MOD Pizza, one of the fastest growing chains in the US**
  "Talk about a great purpose-driven organization, right? They do so many great things for so many great people. SAP runs their ERP and also helps with talent management." MOD's philosophy emphasizes hiring people who are often overlooked and face real barriers to employment, and fosters a sense of opportunity and belonging that spills over into a fantastic, welcoming customer experience. As the company puts it, "But the real secret sauce wouldn't be the product, it would be the people who serve our communities—the MOD Squad. By taking care of employees, they would take care of the customers, and the business would take care of itself."[11]
- **Paradox AI, a conversational chat recruitment agent**
  "They do a lot of work with cultivating networks of talent and ensuring that prospective employees have a terrific experience with that brand throughout the recruitment process."[12]

- **BORN, a global leader in creating market networks**
  BORN-WAVE has revolutionized the way retailers discover the next hot brands. By making the fashion buyer discovery process beautiful, creatives make real connections to the buyers and showcase their brands in the way they were meant to be seen while making fashion B2B ordering simple. SAP BTP seamlessly connects BORN-WAVE buyers and brands to ordering on Knack Systems SeasonOne to realize the BORN mantra: 'Who says B2B can't be beautiful?'

- **The San Francisco 49ers, National Football League team**
  "They wanted visibility into what was happening in the retail shops in the stadium, what was happening with the concessions and in the food court, and what was happening in the premium food area where the season ticketholders were. We worked with them to create a dashboard so they could actually see everything in real-time and worked with them to figure out the right signaling mechanism. In certain parts of the stadium, if I have five incidents or complaints in 15 minutes, I'm going to send somebody there to see what's going on. In another area, the trigger might be five within one minute. They knew how to do that because they knew exactly what their fans were consuming. It is an example of really understanding their customer base and how to view the stadium and the game from a fan perspective (home team and guests). This opens new insights to optimize the environment for compelling fan experiences.

  SAP and the 49ers launched a connected stadium system, called Executive Huddle, designed to help improve how the 49ers capture, report, and respond to game day operations at Levi's Stadium. Team officials can see and monitor real-time data visualizations from nine data sources, including attendance, food and beverage, retail, weather, ticketing, and social media."

Now, let's look at two very different businesses and how the COVID-19 pandemic affected them and influenced their customer pathways.

4.7 The Fressnapf Story

Fressnapf Tiernahrungs GmbH is the largest European pet product retailer. Founded in 1990, it has more than 1,400 stores (many franchised) in 12 European countries. Outside of Germany, it is branded as Maxi Zoo in many countries.

Manuel Cranz, vice president of enterprise architecture, provided more details about the company and its growth. He told us that dogs and cats (and their owners) are its biggest customer segments, but Fressnapf also serves rabbits, birds, fish, and other pets. Stores come in two formats: The normal store carries 8,000 products, and the XXL store offers more than 15,000 products, including aquariums for fish and terrariums for reptiles. Private food and toy brands make up over half of sales.

Founder and CEO Torsten Toeller explained the impact of COVID-19 on Fressnapf. The company was classified as a "system-relevant" retailer, a category similar to "critical industries" or "essential workers" in some jurisdictions. Like most retailers, it had supply chain and panic buying issues. More people working from home sought the company of pets and became customers, creating demand spikes for accessories and other non-food items. Fressnapf stores remained open during lockdowns, and its online business thrived.[13]

Cranz outlined Fressnapf's plans for a blended physical and digital world. In both areas, Cranz foresaw an expanded footprint for the pet supplies company:

> "*Our vision is to build an ecosystem around the pet, providing products and services for our pet lovers and their pets everywhere and in a unique way. The ecosystem consists of the physical and digital worlds. The physical world looks at our stores, including a planned City format for city centers. This area is complemented in our ecosystem by physical services such as grooming and veterinarians. Other services such as dog boarding, dog sitters and breeders, and animal shelters are being planned.*
>
> *In addition to our online stores and customer app, the digital world includes the Marketplace for Pets, where additional pet products are offered to our customers to complement our product assortment. We also offer a digital vet—telemedicine for your pet. Complementary pet insurance can be brokered. We sell our own dog tracker and an associated tracker app that follows a dog's movement and vital signs. Should the dog run away, it can be located through the tracking service in the app."*[14]

Fressnapf is well on its way to omnichannel and work-from-anywhere support. It uses mobile devices equipped with an in-house developed digital assistant to facilitate productivity while giving staff more flexibility away from the office.[15]

Fressnapf's IT landscape has a number of components at the ecosystem level that directly interact with customers (the online shop, mobile app, and Dr. Fressnapf) and stores (the digital assistant, ship from store and order service capabilities, and digital signage). On the core systems level, this includes a host of SAP solutions. In addition to the SAP Retail suite, it runs SAP Customer Activity Repository (CAR) for point-of-sale data management, SAP Commerce Cloud for e-commerce, identity management through SAP Identity Management, analytics using SAP Business Warehouse (SAP BW), and human resources through SAP ERP HCM and SAP Customer Data Cloud. Underlying this architecture is a multi-cloud approach with SAP Business Technology Platform (SAP BTP) and Microsoft Azure. On these two cloud platforms, Fressnapf is developing a microservice and application programming interface (API) architecture to decouple ecosystem access from its core systems.

Taken together, Fressnapf targets the following operational benefits:

- Decoupling the customer and store ecosystem from the Fressnapf core system, which should be only minimally adapted (following the "Clean Core" paradigm)
- Reaching the prototype phase and market faster with new products
- Creating and using the development environment independently, and leveraging automated provisioning there
- Deploying standard monitoring and alerting functionality
- Facilitating continuous development of the provision of data and functions through the domain concept

4.8 Engaging with Customers on the Move

In contrast to Fressnapf's growth during the COVID-19 pandemic, retailers depending on traffic at airports and other travel venues would experience a very different range of challenges. As we discussed in Chapter 3 on the Integrated Mobility megatrend, Gebr. Heinemann is the market leader in Europe and one of the top global players in the duty-free and travel retail industry worldwide.

Roland Vorderwülbecke, director of IT at Gebr. Heinemann, outlined the company's footprint as follows:

- Group turnover of €2.1 billion in 2021
- Up to 700,000 sales items picked daily from two logistics centers
- Supply of around 50 airlines
- Wholesale and retail operations in more than 90 countries
- Supply or operation of around 140 international airports, 200 border shops, and 240 cruise ships and ferries
- Operation of more than 500 of its own retail shops at airports, border crossings, and cruise ships

But operating in the travel industry, particularly in the air travel segment, has brought challenges to Gebr. Heinemann during COVID-19. As Vorderwülbecke explained, Gebr. Heinemann has to work with a variety of different stakeholders with their own agendas—starting with the airports themselves:

> *Most airports are owned or operated by publicly owned companies. We have to win a competitive tender process to qualify for the concession. Next, you have the airlines, many of whom sell their own duty-free items in-flight. Then you have additional stakeholders like booking platforms and travel agencies.*
>
> *Many of our concessions are joint ventures and other partnerships. As a result, we are sometimes a retailer, sometimes only a wholesaler, and sometimes*

something in between—which means we have a really fluid situation. You could position every one of our global locations on a different point on that scale.

Vorderwülbecke noted that his industry occupies a unique niche that also creates branding challenges:

> *There are high barriers to enter the market and there is no global player or household name like Amazon. Most consumers would not even know the names of the biggest players in this sector, like Dufry in Switzerland and Lotte in South Korea. Same with us, and we are number five in the world.*

Part of the challenge is that the company has very few touchpoints with individual customers. Customers enter one of its Duty Free & Travel Value stores only when they are traveling, and only when circumstances allow time for shopping either at the origin airport or at the destination airport. It takes a long time—sometimes years—to develop even two or three touchpoints with a single customer, depending on their travel patterns. How do you convince such customers to share data to participate in a loyalty program and how do you build a community around them? Often, the promotions offered do not fit with travel plans.

The company is moving from a push to a pull model and developing a "next-best offer" mindset. In the past, Gebr. Heinemann has tried out digital displays in stores—one was internally called a "Digital Instore Assistant" to guide customers—but utilization was low. Now, the company is leaning more towards a high-touch clientele service model in their stores, according to Vorderwülbecke:

> *We are starting to think more about enhancing our store sales staff. We'll provide easy checkout for customers who prefer convenience. They know what they want and prefer to leave as quickly as possible. On the other hand, we want a different approach to customers who like to browse and are open to conversations with our salespeople, who can tell them about heritage of specific wines or give skincare advice around cosmetics.*

For Vorderwülbecke, Gebr. Heinemann's IT landscape has four main layers, as shown in Figure 4.2:

- Its travel retail core is Gebr. Heinemann's IT backbone. This area covers functionalities to sell wholesale and thus supports related supply chains. These are mostly SAP solutions run in a private cloud environment. This core is very much focused on cost efficiency and reliability.

- The commerce solution is more about customer-centricity and personalization, and the response to customers' needs in their blended B2B and B2C models.
- The service platform connects the customer side—the traveler—to the retail core.
- The fourth layer is about everything they need to make Gebr. Heinemann's employees more productive: tools and access to all capabilities.

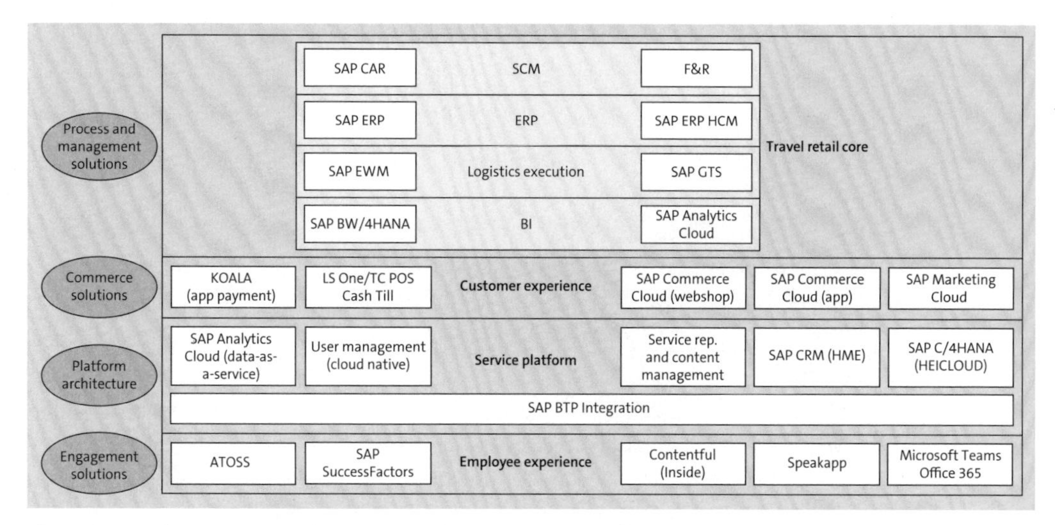

Figure 4.2 Gebr. Heinemann's Current Solution Landscape (Source: Gebr. Heinemann)

Over 6,000 kilometers away from Gebr. Heinemann's headquarters, the Adani Group is also rethinking the airport experience. As one of India's largest conglomerates, the company is poised to digitize and optimize the airport experience for travelers. The Indian government has been privatizing many airports, and the Adani Group now operates and maintains airports, making them much more digital for new traveler experiences.

After meeting with the Adani Group, Bhargava of SAP described the organization as *jugaad*, a word that has become popular in Indian business circles for someone who is resourceful at delivering frugal innovation. The Adani Group uses SAP's commerce functionality, continuously challenging us to enable creative new ways to engage with travelers and shape convenient customer journeys with "buy now, pay later" options, contactless checkouts, dynamic pricing, or personalized promotions.

Bhargava explained that airport operations is only one part of the Adani Group's portfolio. As a conglomerate, the company is engaged in renewable energy, mining, natural gas, food processing, retail, and infrastructure operations:

> *What impresses me about Adani is that they're investing in digital transformation as a growth lever, not only as an efficiency driver. That's a paradigm shift for me in India, as businesses are thinking about digitizing compelling experiences and not just optimizing for selling products and commodities.*

4.9 The Digital Commerce Journey Accelerates

It is fascinating to see how far the retail sector has come and how it defends its place at the forefront of digital transformation. In his book *Silicon Collar*, industry analyst Vinnie Mirchandani describes the evolution of scanning technology in stores.[16] Figure 4.3 shows the original UPC code as a bullseye shape, as patented by Norman J. Woodland and Bernard Silver to create a unique apparatus and method for classifying items.[17]

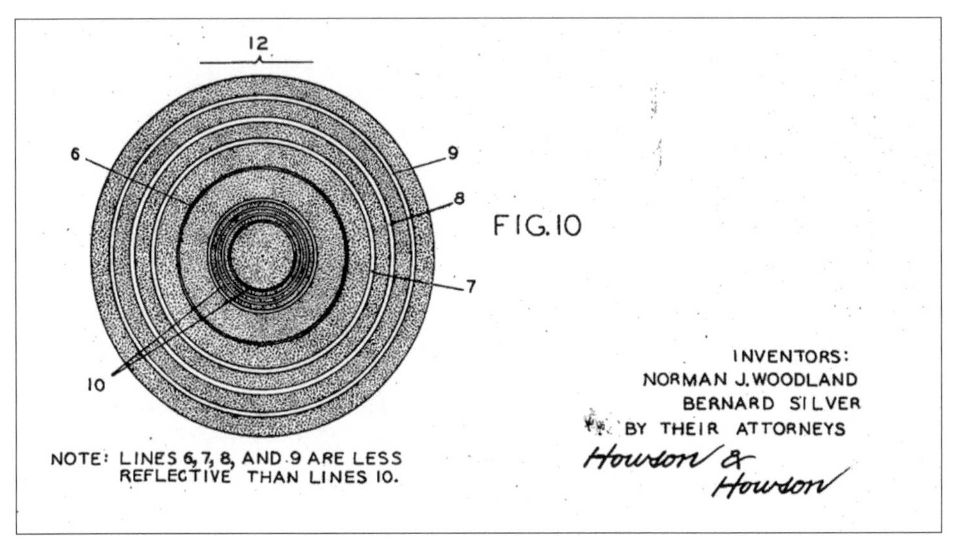

Figure 4.3 Original Universal Product Code (US Patent 2612994)

Consider the rectangular barcode on the packaging of virtually every product you pick up in a store. Although patented in 1952, scanner technology wasn't available until the 1990s, when scanners in stores became more common. Four decades! Today, technology innovation and adoption in retail stores and supply networks occur much faster, accelerated by reams of data generated by myriad sensors and cameras that track the behavior of digitally connected consumers. AI tries to make sense of the data to drive a customer-centric evolution and transformation.

Bhargava described how the SAP architecture supports its customers. The management of commerce, service, and customer data is the technical and business foundation, provided by SAP S/4HANA. The next layer enables different business models, from B2B over B2C to all the variants of business-to-business-to-consumer (B2B2C).

> *Across CX, there needs to be a layer that breaks down the process- and data-based organizational and technical silos. This flexible and modular layer provides the basis for a great deal of business innovation. For example, our intelligent commerce solution will leverage this layer in the future. Our customers experiment a lot and not everything will work out, stick, and stay, so we give*

them the option to subscribe to modular extensions on a consumption basis. This way they pay 'by the drink' on their journey to shape the ultimate customer experience.

SAP has been firmly in the B2B business for five decades, and now B2C practices and preferences are catching up with the way we engage with our customers across all industries. SAP offers an innovation space for experimental models that are ready to scale when an idea takes off.

4.10 New Customer Pathways with SAP

Before we close this chapter on new customer pathways, we want to highlight some of the solutions in SAP's portfolio that offer key functionality for managing customer data, managing commerce activities and engaging with customers, planning and optimizing product assortments, managing omnichannel promotion pricing, orchestrating and managing orders, handling customer returns, and improving sustainability.

4.10.1 Customer Data Management

Customer data management connects information from across your enterprise to inform business decisions, build trust, and strengthen loyalty while respecting your customers' data privacy and reducing your compliance risks.

The following solutions contribute to customer data management:

- SAP Customer Data Platform analyzes customer behavior and demands, offers better insights, and tracks customer activities and segmenting data to provide personalized experiences.
- SAP Customer Data Cloud enables customers to control their profiles, preferences, and consent data; supports customer profile management by transforming customer identity, profile, and account status into a unified customer profile; and creates a frictionless experience across brands, channels, and devices using secure registration and social authentication to identify online visitors from any touchpoint.

4.10.2 Commerce Management and Customer Engagement

Commerce management and customer engagement converts browsers into buyers by providing exceptional and personalized experiences throughout every interaction. The following solutions contribute to commerce management and customer engagement:

- SAP Commerce Cloud can engage and transact with customers across all channels and touchpoints consistently and deliver rich and engaging information to drive conversion.
- SAP Emarsys Customer Engagement takes customer experiences to the next level with real-time, personalized omnichannel engagement.
- SAP Omnichannel Promotion Pricing helps companies offer a true omnichannel experience to their customers by enabling them to provide consistent promotion pricing across channels.
- SAP Intelligent Returns Management guides products from consumers' hands to the final dispositioning steps at the end of the product's lifecycle, maximizing both customer experience and margins.

4.10.3 Assortment Planning and Optimization

Assortment planning and optimization puts the customer at the center of all merchandising decisions with curated product mixes and hyperlocal assortments of sustainable products. The following solutions contribute to assortment planning and optimization:

- SAP Assortment Planning allows retail operations to select and plan product mixes for stores based on consumer buying behavior.
- SAP Customer Order Sourcing helps enterprises make intelligent order sourcing decisions based on their business requirements to deliver a great buying experience across sales channels, by providing centralized access to availability information from all locations in all channels.

4.10.4 Omnichannel Promotion Pricing

Omnichannel promotion pricing ensures effective sales prices across all sales channels along with the ability to introduce new promotion rule types with low implementation effort. SAP Omnichannel Promotion Pricing helps companies offer a true omnichannel experience to their customers by enabling them to provide consistent promotion pricing across channels.

4.10.5 Order Orchestration and Management

Order orchestration and management involves streamlining commerce operations by processing, controlling, and routing orders from any touch point, as well as allowing buyers to get an instant view of their recent and current orders. SAP Customer Order Sourcing helps enterprises make intelligent order sourcing decisions based on their business requirements; this way they can deliver a great buying experience across sales channels, by providing centralized access to availability information from all locations in all channels.

4.10.6 Sustainability in Customer Relationships

You can drive sustainability at scale by embedding operations, experiences, and financial insights into your core business processes. Feather by SAP powers the business processes required for managing the *re-commerce business* in house. This area includes backend integrations and workflows required to add and track inventory, oversee product movements, post items onto a branded storefront, and feed transaction data into reporting dashboards.

Chapter 5
Lifelong Health

The COVID-19 Catalyst

We're writing this book during a global health and well-being crisis: the ongoing COVID-19 pandemic has directly and indirectly impacted millions of people around the world. We have a healthy tendency to focus on humanity's ability to adapt even during crisis, but let us not forget the toll of the pandemic: As of late 2022, more than 6.5 million people have died from or with COVID-19. If you include the impact on families and friends, the virus has directly brought pain and suffering to over 10 times this number of people.

The last few years have been sobering: As we reflect on the challenges we've faced, we must also look ahead to how these challenges and their solutions can be applied in the years ahead. We've seen exciting advancements in many spaces. The clinical trials for the COVID-19 mRNA vaccines were some of the quickest ever recorded. The logistics of global vaccine distribution were complex, demanding cold-chain specifications, and many distributors came through with flying colors. Applying machine learning to mRNA biology is leading to potential new applications and drug candidates. Telemedicine took off. Hospitals learned to reconfigure their facilities and workforces at breakneck speed. The lack of previous investment in public health highlighted a need for correction. Mental health received a lot more focus. Many companies elevated the role of the chief medical officer (CMO). Ordinary citizens became much more medically aware thanks to COVID-19-tracking dashboards like the one at the Johns Hopkins University School of Medicine.[1] Focus on value for money has grown, increasing to attention to the many inefficiencies found in healthcare systems, especially in the US.

In its 2021 collection of articles, *McKinsey on Healthcare: Perspectives on the Pandemic*, a McKinsey analyst summarized: "During the COVID-19 crisis, one area that has seen tremendous growth is digitization, meaning everything from online customer service to remote working to supply-chain reinvention to the use of artificial intelligence (AI) and machine learning to improve operations. Healthcare, too, has changed substantially, with telehealth and biopharma coming into their own."[2] We begin our discussion of the Lifelong Health megatrend with our recent experience with COVID-19 in mind.

Healthcare Analytics

Let's start with a perspective from Michael Byczkowski, SAP's global vice president and head of the healthcare industry business unit. While healthcare is a highly local affair due to myriad regional rules and certifications, Byczkowski brings a global perspective when dealing with healthcare customers and partners around the globe—whether in Australia, Canada, Germany, Singapore, South Africa, the UK, the US, and beyond.

Byczkowski provided many perspectives on the extraordinary effort required to react to COVID-19. He started with Heidelberg University Hospital (Universitätsklinikum Heidelberg, UKHD) as an example of digitization as the backbone for data exchanges and for running processes. UKHD, he said, has over 40 specialist departments, including oncology, that offer world-class medical care and attract patients from around the world. The hospital's active collaboration with national research facilities has produced several Nobel Prize winners. During the pandemic, Byczkowski noted, the hospital was tapped for emergency response:

> *The challenge was that the manual processes in place for routing patients would not be able to handle the large influx of emergency patients with COVID-19 requiring isolation and other precautions. Medical staff needed a way not only to see an overview of ICU beds on the fly, but also to get an inventory of all available beds per ward for COVID-19 and non-COVID-19 cases. This included whether these beds were readily available, meaning that they were equipped with needed devices like ventilators, for example.*

According to Byczkowski, the key for UKHD could be found in high-powered analytics applications providing new insights into the hospital's patients, processes, and data, as shown in Figure 5.1:

> *The solution was a rapid development on the SAP Business Technology Platform all within a week of a complete solution, including a portal that provides easy access to manage the underlying data and guide the right decisions. Two applications manage the registry of roles: one for managers and workers in hospitals who can see and edit data from their hospital, and one for reporting viewers who can see data across various hospitals and wards. Other applications let managers maintain data for hospitals and wards and confirm onboarding requests for new users. SAP Analytics Cloud supports applications that aggregated the data for each hospital and displayed it according to hospital-specific requirements. Four other analytic applications deliver additional insights like geo data and forecasts.*

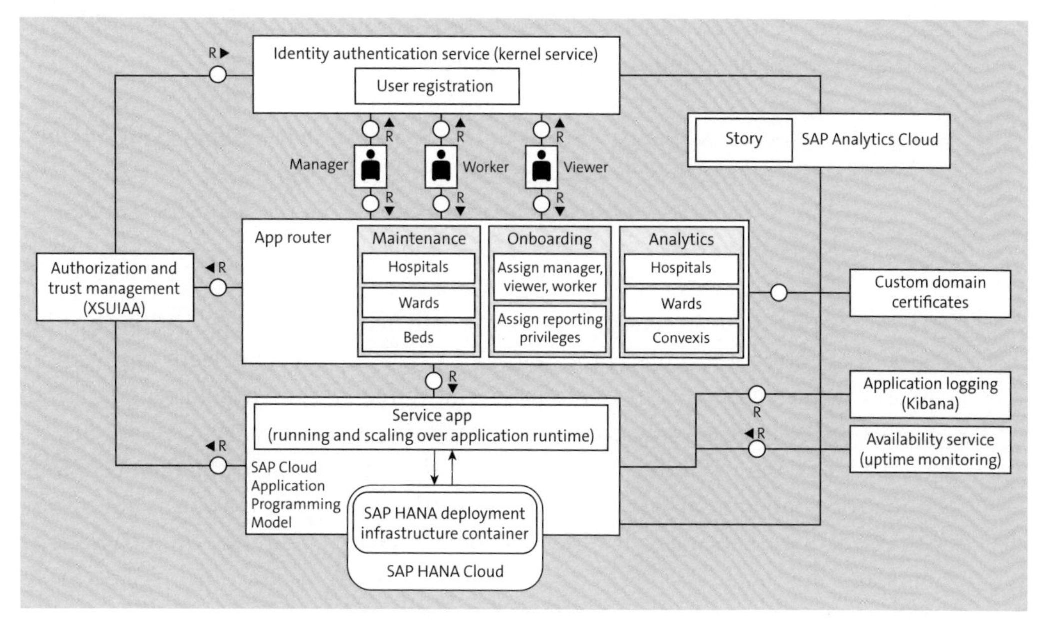

Figure 5.1 IT Architecture at UKHD (Source: UKHD)

Byczkowski highlighted another key gain that came from the creation of a coordinated service center that facilitated the whole data-centric system:

> *The COVID-19 coordination center at UKHD allows doctors to control and manage all the information they need to coordinate patient care and transportation more effectively during the pandemic. The local health authority could also provide real-time reporting to the state and federal authorities about the availability of COVID-19 hospital beds with and without ventilation options.*

Workforce Resilience

Byczkowski then turned to how workforce resilience, sustainability, and mental health became crucial because of the frequency of mutation of the COVID-19 virus, which brought waves of variants and thus surges of new patients and uncertainty about whether previous treatments and protocols would continue to be effective.

He focused on Coppelia Rose, global human experience management and healthcare leader at DXC Technology in Brisbane, Australia. DXC is a multinational SAP partner, working with SAP across several industries and countries. Rose represents a division that has worked with over 1,600 healthcare customers around the world. This division has significant industry credibility earned partly through a group that developed provider-side software. In 2021, that unit was sold to the healthcare systems company Dedalus Group, based in Italy.

Rose elaborated on how DXC responded to the needs of the day when COVID-19 hit by bearing down on the useful and the readily available:

> *We had been working with SAP SuccessFactors talent management modules in our healthcare offering. With our healthcare experience, we could allow customers to hit the ground running with some pre-configurations and other tailored content.*

DXC worked with customers in a pandemic-fatigued industry that had pushed big digital transformation projects to the back burner. These clients were not ready to discuss the introduction of new enterprise-wide human resources solutions or payroll systems, so DXC focused on projects with a short time to value, such as helping healthcare organizations sift through filing cabinets to find out which staff members were certified or competent to handle communicable diseases. What it saw in its direct work with essential workers was a growing need for mental health support; the long working hours and the many lost lives took a toll on the staff. As described by Rose, DXC designed a solution:

> *We specifically saw an opportunity to help frontline healthcare workers and developed an eight-module Workforce Suite for Healthcare (available on the SAP Store). What we saw with the frontline workers was stress and burnout. So, we said, 'What could we do to support that?' It's all well and good to measure it. We can do a survey and measure stress. But what are we offering to actually help on the ground?*

DXC developed a mental health support app that complemented the solution. This mobile application for frontline healthcare workers allows them to connect with other people in a similar role. Working long hours in protective gear fighting for (and often losing) the lives of patients created extreme stresses for frontline workers, and connecting with peers to talk about their experiences and struggles was important:

> *I was interested to learn about the moral injury associated with healthcare workers where they ask, 'Could I have done more? That patient didn't get to say goodbye to their family, and I was standing there with an iPad.' It's different from trauma. Exposure to trauma is, 'I've seen patients that have lost limbs.' But with COVID they suffered from what is called moral injury, and that's a different thing to cope with.*

The solution uses features from SAP SuccessFactors. Then, DXC brought in SAP Fieldglass capabilities for contingent workers, travel nurses, agency nurses, and *locum tenens* (temporary) physicians. (*Locum tenens* is Latin for "placeholder.") Next, they brought in Qualtrics XM for the employee experience and extended that product to incorporate an Oldenburg Burnout Inventory calculator, which runs on SAP Business Technology Platform (SAP BTP), to measure occupational burnout and stress. The first-line value from DXC's mental health support application is help for frontline health-care workers in dealing with stress and moral injury, which in turn is also good for patients and medical outcomes. The solution also reflects that the patient experience is an important metric today, particularly in the US with its highly privatized healthcare. Providers may not get paid if they receive unsatisfactory patient experience scores:

> *If you have a very stressed-out clinician who has worked too many shifts and whose bedside manner has suffered as a result, you'll get a low patient experience score, and in some countries the reimbursement from health insurance providers may be impacted.*

For Rose, a case management component to systematically engage with frontline medical workers who suffer from pandemic-induced mental health issues would be highly useful:

> *We could easily have added a ninth module for case management. Now, that's not about a human resources ticket management. This is case management as in return to work from bereavement or long-term stress leave, with mental health professionals and HR managers and general practitioners or other doctors having access to an employee's record to actually manage a return-to-work case after a long-term stress period.*

The need for healthcare has grown because of the pandemic, but at the same time, Rose said, the system must deal with a workforce population that has reached the end of its rope:

> *They are already so stressed. So, the minute they have a loss in their own family that maybe has nothing to do with COVID, they're already depleted from every bit of mental fortitude they have and end up on stress leave. So, then there's a return-to-work program. More and more I'm seeing a demand for it. It's specific to the industry because of the type of injuries. It's not just PTSD.*
>
> *The term that I've heard used recently, by Dr. Lisa Bellini from Pennsylvania University Medical Center, was, 'Our teams are languishing ... not caring anymore.' It's no longer stress and burnout. It's just, 'Ugh! I just can't. I just can't do it.'*

Rose brought home the point that, for companies to retain (let alone gain) their essential frontline workers, they definitely need a mental health software component for their staff, especially in the US and UK:

> *And so, that's why we've put a lot of focus on the mental health support side, and we're working to implement more than just the business systems. The business systems help the organization be an employer of choice. But there still needs to be something in the hands of the end users.*

DXC found that the UK and the US had the highest stress levels, and many have heard about the "Great Resignation" resulting from the pandemic. Rose noted that in the US, 92% of registered nurses who were surveyed said they were actively thinking of leaving the profession:

> *Ninety-two percent! And we have been able to say, 'If you could show that you were an employer of choice, if you could offer succession and retention incentives and performance metrics and active learning and support for mental health and really care about your employees, you may retain more people.'*

SAP is also in an industry that fully depends on an engaged and motivated workforce. A report that would even remotely show these kinds of numbers would send shockwaves through our executive board, so we can imagine that workforce health and retention are on the top of the agenda in each hospital. Without frontline health workers, high-tech intensive care units (ICUs) would be vacant, resulting in more people dying and in more hospitals going out of business or barely hanging on.

The Vaccine Collaboration Hub

Byczkowski next turned to the challenge faced by the life sciences sector after the mad rush for effective vaccines resulted in the breakthrough successes of mRNA vaccines from Pfizer/BioNTech and Moderna. Holding some vaccine in deep-freeze storage in a factory is one thing. Having hundreds of millions of people lining up for shots around the world is a different scenario:

> *In the beginning, we found it was not a matter of supply but of demand, something that is absolutely not unheard of. Think of Apple and how they managed supply for their incredible demand of memory chips prior to launching the first iPod with a solid-state drive. There was so much more demand than supply was available. We needed to network everyone, including the pharma companies, the wholesalers, the distributors, the healthcare providers, government health*

agencies, pharmacies, vaccination centers. This is the story about our Vaccine Collaboration Hub and how everything came together.

SAP already had the Information Collaboration Hub (ICH) for Life Sciences to help the industry in its elaborate track-and-trace requirements to fight counterfeit drugs. The solution creates a network of marketing authorization holders (comparable to original equipment manufacturers in the automotive and other discrete manufacturing industries), contract manufacturing organizations, providers of active pharmaceutical ingredients and other substances, third-party logistics providers, and regulatory bodies. The ICH can exchange large volumes of data and messages in different formats to meet quality management and regulatory requirements. The Vaccine Collaboration Hub shown in Figure 5.2 extended the capabilities of the ICH to meet pandemic-specific requirements.

> *The Vaccine Hub, along with SAP Integrated Business Planning for the supply chain and SAP S/4HANA, helped the German Red Cross of Saxony. Managing multiple doses of vaccine injections in a timely manner was challenging. They simulated, planned, and managed the flow of vaccines from regional distribution centers to vaccination centers and, ultimately, to citizens. Lars Werthmann, head of logistics at DRK, said, 'Humanitarian logistics is a big challenge in our fight against the pandemic.'*

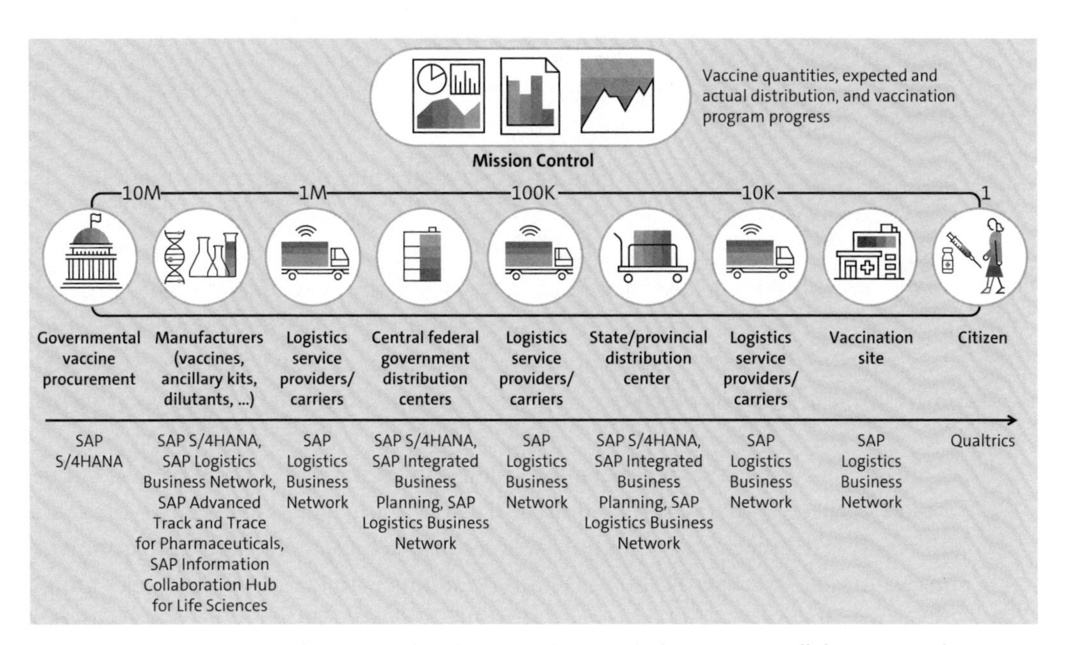

Figure 5.2 Visibility across the Demand and Supply Chain with the Vaccine Collaboration Hub (Source: SAP)

Informed Citizens Fight against the Pandemic

A nasty virus stays quiet for a few days before symptoms alert us to its arrival. A really nasty virus like SARS-CoV-2 is highly contagious during this asymptomatic period. Governments faced the challenge to trace contacts and warn people if they had contact with infected people so they could isolate and protect others. Byczkowski explained:

> *SAP and Deutsche Telekom developed and rolled out the German Corona-Warn App, which runs on Apple iOS and Google Android platform. The app was developed as open-source, and the program code has continuously been accessible to the public on the GitHub platform.[3] It was rapidly developed in a matter of weeks and was first released in June 2020. By autumn 2022, it had been downloaded more than 47 million times and is available in many languages, even Ukrainian.*

This project was quite an experience for all of us, from the executive board to our experts who worked with Android and iOS experts to provide application programming interfaces (APIs) to enable our app to access to the signal strength measured by the Bluetooth module when your smartphone comes near other smartphones. This creative and technically interesting idea used signal strength as a coarse but very practical way to measure distances between people, but our true challenges were quite different: We at SAP don't usually build software for consumers; we don't usually build software with politicians, activists, data privacy experts, journalists, and citizens breathing down our necks; and we don't usually build open-source software and certainly not in a few weeks. But the team pulled it off, and we are very grateful for their efforts and success.

Byczkowski commented on the pandemic as a catalyst for broad transformation:

> *COVID-19 accelerated digitalization in many aspects of healthcare, but it was inconsistent. By having to treat so many infected and infectious patients, hospitals clearly had other priorities to attend to first. It's the same topic regarding public health agencies. There is also the need to take the impressive volume of medical research and learnings and translate it into clinical routine.*

COVID-19 was definitely an accelerator. Herculean efforts by so many people, unprepared for such a scenario, were required to deal with the pandemic. Plato's aphorism "Necessity is the mother of invention" was confirmed once more. In this environment, innovators like Byczkowski thrive. He was brimming with ideas and challenges for his colleagues, customers, and partners: "What did we learn and what do we do with our learnings?"

5.1 The Cutting Edges in Healthcare

While perhaps still too early to wave the COVID-19 pandemic goodbye, it's not too early to review what we've learned from the first three years of the pandemic: How can we be better prepared for another public health crisis of a similar magnitude, and what role can digital technologies play in this space?

Byczkowski highlighted four promising technology trends: advances in access to personal health records, the use of AI in healthcare settings, the proliferation of telehealth services and wearable technology, and genomics and personalized medicine.

5.1.1 Personal Health Records

For years now, doctors have been warning us about "silent killers"—diseases like hypertension, heart diseases, or diabetes that can go unnoticed for a long time but can have fatal potential. More recently, researchers at Johns Hopkins University found that "medical errors [are] now third leading cause of death in the U.S."[4] Dr. Martin Makary noted that "the medical coding system was designed to maximize billing for physician services, not to collect national health statistics, as it is currently being used."

Byczkowski said:

> *We knew it even before COVID, but the pandemic has made us realize how important it is to have a comprehensive and portable patient record. Today, the data is fragmented across multiple providers, across claims in payer systems, across clinical trials conducted by pharma companies, and across databases in public health agencies. The patient's fitness tracker and imaging records are rarely readily available at the point of care. The healthcare systems were designed to be hospital-centric and insurance company-centric, not patient-centric.*

He has convinced us that the solution is not a single super-database with all patient information. What we rather need is an approach to connect patient data in all those distributed systems at public and private hospitals, general practitioners, and laboratories. Access would be granted upon approved request. Byczkowski provided an example of how even an isolated database can provide significant benefits:

> *We developed a cancer database for a university hospital in Germany. It stores all the patient data for cancer treatments going back nearly 30 years. You can check the exact diagnosis, the exact treatment, whether it was a systemic therapy with different pharmaceutical products, surgery, radiation therapy, or any combination. Over time, you get a great overview of what worked best*

for which diagnosis. Then you have highly valuable data that gets even more valuable over time, since the hospital also performs regular follow-ups and check-ins as long as the patient lives, even when they move to different locations or countries.

The hospital definitely took a thorough approach to lose as few patients as possible in this long-term study. Even if a patient moved from Germany to Mexico, that patient would still be called twice a year for an update of their current health status. Byczkowski has cautioned that this particular database was super-focused on cancer treatment. But you can still imagine what a difference it would make for doctors and patients if we could someday create a composite image of patient data, similar to what Discovery Health in South Africa can assemble. Naturally, some eyebrows will be raised when an all-encompassing health database is proposed, even if that database is distributed. For Byczkowski, a privacy concern is on many people's minds:

“ *Today, there are plenty of privacy concerns and competing vendor restrictions that make such data consolidation difficult. In early 2021, the largest digitization project in the German healthcare system was launched: the electronic patient health record. It took 16 years of preparation!*

Mandar Paralkar, global vice president and head of the life sciences industry business unit at SAP, said the technology industry has long had protocols to safely exchange patient data, but these protocols are often regional. With the caveat that the same may not apply to other countries, he commented from a US perspective:

“ *If a healthcare-relevant application is storing and processing data, IT vendors need to ensure—and provide evidence—that their software fulfills all regulatory standards. If an IT vendor works with subcontractors in India or Brazil, for example, then they need to ensure that business associate agreements are in place with subcontractors. IT vendors should have those back-to-back agreements if the applications are operated in the data center of a hyperscaler.*

Byczkowski commented on the importance of trusted data custody:

“ *Our core advantage at SAP is that we are trusted as being a kind of 'Switzerland of data.' We have a solid reputation across all industries as a good custodian of our customers' processes and data. We don't want to look at their data or mine them for secondary use cases. It's like when you rent a safe deposit box in a bank vault: what you put in the box is not our business.*

SAP also doesn't own our own clinical information system, so we are an attractive partner for health system vendors and healthcare providers. Because we are neutral, we can connect companies seeking to derive additional value from their data in the data economy.

5.1.2 Artificial Intelligence

Byczkowski co-authored "The Industrialization of Intelligence," a chapter in an anthology that examines how humans have acquired medical knowledge over the centuries, handed down over generations, augmented with observation and experiment.[5] Today, computer imaging and genomics analyze the digital code that programs our cells and bodies. Patient data and insights into diagnostics and therapies grow along with computing power; AI in its various manifestations has shown up in pockets along the healthcare spectrum. In its briefing, *Transforming Healthcare with AI: The Impact on the Workforce and Organizations*, the consulting firm McKinsey & Company outlined a few key areas, such as clinical support, care delivery, chronic care management, self-care and wellness, triage and diagnosis, and diagnostics.[6]

Byczkowski commented on AI applications:

> *Natural language processing [NLP] helps doctors transcribe their voice recordings, finally making doctors' scrawl readable and processable. IBM Watson had some other initial successes.[7] But what if we could support a tumor board in diagnosing and creating the treatment plan by having doctors from the different disciplines look jointly at the patient data. You might say, 'This is state of the art.' But what if we could show the board the history of patients with similar symptoms and diagnoses, how they were treated, how they responded to different treatments, and what their outcomes were? I think this would tremendously help determine the best treatment path for this patient. And that's where the current discussion is heading.*

In "The Industrialization of Intelligence," Byczkowski cited the example of British ship doctor James Lind who conducted "the first controlled comparative study in the history of medicine in 1747, identifying the use of citrus fruits as an effective way of treating scurvy." He also used a modern example how far we have come in another paper:[8]

> *The AI-driven Ping An Good Doctor online platform in China has more than 400 million registered users for services like online consultations, referrals, registrations, and online drug purchases. They have an in-house medical team*

with more than 1,800 experts. The Good Doctor knows 3,000 diseases and is continuously trained with data from hundreds of millions of consultations. The AI system doubles the efficiency of doctor consultations, and greatly reduces the number of misdiagnoses and missed diagnoses. This type of high-tech diagnostics is available in the most remote Chinese village and improves the patient's experience with remote medical consultations.

5.1.3 Wearable Tech and Telehealth

After decades of dawdling, telemedicine took off during the pandemic. But Byczkowski expects much more from this technology than just supporting outpatient consultations over Zoom. The prerequisites are mostly fulfilled already: 500 million smartwatches and other wearables contain more sensors and AI than your average ICU had only a few decades ago. Diabetes patients have implanted sensors and insulin pumps. Intelligent pacemakers and implanted defibrillators save many lives. We don't even need breakthrough technologies to make a big difference for people's lives if we integrate the sensor data we accumulate while we are awake and sleeping into our medical histories. Even if we prefer to see our doctors in person, they can provide more accurate diagnoses and treatment plans based on more and better data.

But that will look primitive compared to scenarios that some foresee for 2050: a future where showers are outfitted with interfaces and sensors that can perceive personal health metrics, transfer the data to a care team in real time, and receive via taxidermal infuser the physician-recommended dose of vitamins, pain relievers, and prescription medications.[9]

Admittedly, this 2050 vision feels scary and maybe even completely delusional. On the other hand, think back 30 years to when the internet was in its infancy, cell phones were a new thing, and smart phones were science fiction. So, we suggest you bookmark this page and come back in 30 years to check how close the high-tech shower scenario was to our future reality.

Byczkowski had a more grounded outlook on trends in the healthcare industry:

- **Smart hospital**

> *Integrated platforms are using digital documentation to create actionable insights for the entire organization. Smart hospitals use data for treatments and also for managing and optimizing administrative processes. The SAP portfolio delivers the capabilities to support the business functions of a healthcare provider including analytics, human resources management, finance, asset management, or supply chain. SAP's process management tools can optimize existing processes on the journey to becoming a smart hospital.*

- Patient centricity

> *The patient is becoming a partner in healthcare using mobile devices and healthcare gadgets like activity trackers to monitor their own fitness and well-being. The patient should be at the center of all activities and not be just an object that you're treating. We are seeing that even more as patients provide their own data, from fitness trackers or medical imaging that was done somewhere else. And educated by Dr. Google, patients play a more active role in diagnosis and treatment—not always to their or their doctor's advantage.*

- Data-driven decision support

> *Current technologies support medical and administrative staff in their daily work and decision-making processes. Integrating artificial intelligence into data- and image-driven decisions is a significant cultural change in a classical hierarchical culture. We expect human doctors to take their silicon colleagues seriously when they offer a diagnosis that deviates from their own opinion. But mistaking a malignant for a benign tumor (or the other way around) can have disastrous consequences for the patient.*

- Workforce experience

> *A well-trained workforce in healthcare is scarce and top talents must be retained to stay competitive in the market and stand out by delivering best-in-class healthcare services with an engaged workforce.*
>
> *After years of excitement about high tech in the operating theater, the pandemic brought back the harsh reality that well-trained and motivated nurses have a significant and often underestimated impact on patient outcomes. The massive workloads medical staff encountered led to mass resignations as medical staff were overworked and burnt out.*
>
> *Workforce experience is also about employee engagement. Talent retention is critical to stay competitive You can only stand out by delivering best-in-class healthcare services, which brings great patient experience and happy customers as a result.*

- Healthcare networks

> *The exchange of data and knowledge helps create a sense of community with new and innovative ecosystems that support the use of data for better patient outcomes.*

The scientific enterprise—and we count medicine firmly in this domain—relies on the exchange of information to create a system that ultimately delivers better patient outcomes. Patient care and health sciences both operate in a highly networked conglomerate of industries, but not all of the information is available where it is needed. Healthcare providers only have partial information. Payors have claims and payment information. Population health data is fragmented, with public health agencies collecting some of that data. Pharmaceutical companies retain most clinical trial data. So, let's look at the potential of healthcare networks for patients: new and innovative ecosystems that support the meaningful use of data to improve patient outcomes.

5.1.4 Genomics and Personalized Medicine

In 2003, the Human Genome Project gave us "the ability, for the first time, to read nature's complete genetic blueprint for building a human being."[10] Meanwhile, the National Human Genome Research Institute (NHGRI), part of the US National Institutes of Health (NIH), provides this definition:

> *Personalized medicine is an emerging practice of medicine that uses an individual's genetic profile to guide decisions made in regard to the prevention, diagnosis, and treatment of disease. Knowledge of a patient's genetic profile can help doctors select the proper medication or therapy and administer it using the proper dose or regimen. Personalized medicine is being advanced through data from the Human Genome Project.*[11]

Byczkowski said he sees a game changer for patients and the health system:

> *We talk a lot about patient-specific drugs for rare diseases caused by monogenic disorders. There a drug is specifically designed for individual patients and their genetic setup. Defective genes can often be fully repaired, healing the patient completely. This can truly save people's lives—and spare public health systems millions of dollars/euros because one does not need to treat or support the patients over their entire remaining lifetime.*
>
> *But more often one checks for specific biomarkers to ensure that a drug can actually deliver the expected results and not just cause negative side-effects. In breast cancer, for example, a treatment with Olaparib is only viable if the patient has a mutation in her BRCA1 or BRCA2 genes.*

The research is accelerating: In recent years, drugs to treat what are considered "rare" diseases represented 25% of all newly introduced drugs with new active ingredients.

Next, we'll look at a flurry of healthcare and life science innovations, focusing on how two companies from different corners of the world are applying exciting technologies and considering how their experiences with lifelong health innovations can inform your organization's next steps.

5.2 The Discovery Health Story

Let's face it, insurance is the business of statistics and probabilities. In this system, we aren't really customers but "risks" and "premium payers" instead. People with our risk profile have certain probabilities to develop a health condition or crash our car. If this happens, the insurer grudgingly pays for our claims after we have grudgingly paid the premiums over the years.

But we should remember: the insurer knows that perhaps 1 in 100 people will get seriously ill in a year, and we know that getting seriously ill may bankrupt us if we are uninsured. So we should be happy to pay the premium to offset this risk and even happier if we are lucky enough to be among the 99 healthy people.

Discovery Health would challenge this system. This organization was founded in Johannesburg, South Africa, in 1992. One of its founders, Adrian Gore, wrote in *McKinsey Quarterly* about "gut instinct" when he saw the connection between lifestyle choices and sustainable insurance:

> *"When I started out as a young actuary in a life insurance company, South Africa was moving from an apartheid state to a proper democracy and was facing some serious challenges, particularly in healthcare. There was an undersupply of doctors, an unusual combination of disease burdens, and a new regulatory environment that had zero tolerance for the discrimination of the past, and rightly so. This meant you couldn't rate customers on preexisting conditions. Finally, unlike most countries, where a national system partially covers risk, there was no unified public health insurance system at that time.*
>
> *When you put those four things together, sustainably financing healthcare becomes a very complex undertaking. When we formed Discovery, we asked, 'How do you innovate and build a health insurance system that can work in this kind of environment?'*
>
> *Our gut instinct was that if you can make people healthier, you can offer more sustainable insurance. It turns out that three lifestyle choices (smoking, poor nutrition, and poor physical activity) contribute to four conditions (diabetes, cancer, heart disease, and lung disease) that drive over 50% of mortality every year. So, lifestyle choices are fundamental to any social insurance system. The*

behavioral science tells us that people need incentives to make a change. But that wasn't universally known at the time; we were just a start-up acting on a hunch.[12]

After three decades, Gore's vision has become reality:

> *That initial idea was the genesis of our Vitality program, which has evolved into a complete wellness system that tracks everything from physical activity to nutrition over the course of a person's life. For instance, customers earn points by logging their workouts with fitness devices from Nike+, Fitbit, and others. These sync up with Vitality directly, through a computer or with mobile apps on smartphones. When you go to our partner grocery stores, the healthy food is clearly demarcated on the shelf, and you get a 25% discount at the register when you swipe your Vitality card. When we first launched the program, we were criticized for wasting healthcare dollars on incentives, but customers went berserk for it.*[13]

Maia Surmava was chief information officer (CIO) and head of clinical systems at Discovery Health and has recently been appointed CEO of Vitality US, a subsidiary of Discovery Health. She explained her company's multi-faceted profile:

> *We're not your traditional health insurer that measures and prices risk; rather, we manage and reduce risk, delivering on our core purpose of making people healthier.*
>
> *As a payor and care enabler, we play several roles. For example, in the digital health space, we empower our members with health insights and identify preventable risks and best actions to improve their health. We're laser-focused on navigating members to the right care, at the right place, at the right time, and supporting our members at the key points in their healthcare journey. At the same time, we empower providers with insights to personalize treatment and trigger appropriate interventions.*

Surmava expanded on the company's journey to becoming a "digital health enabler":

> *We started in South Africa as a health insurer in 1992, but we have since diversified into life insurance, short-term insurance, investment management, and banking. Twenty-five years ago, we created the category of incentive-based behavior change and wellness programs, leveraging incentives, data, and the latest technology to tip the scale in favor of wellbeing and to reduce over-reliance on sick care. Today we are active in 40 countries, impacting over 30 million lives.*

Surmava explained that Vitality is the world's largest platform for incentivizing and rewarding behavioral change and serves as the engine for Discovery Health's partnerships with life and health insurers. Discovery Health continues to expand: Vitality Health International is a new division that provides international travel insurance and travel packages for healthcare treatment. Amplify Health is a joint venture with AIA Group and brings healthcare and insurance solutions to the Asian market. Discovery Bank claims to be "the world's first behavioral bank," which rewards customers' responsible behavior such as reliably paying bills with better interest rates and other offers.

Surmava spoke about how the model of influencing the risks with behavioral incentives can be scaled through the numerous partnerships Discovery Health is shaping around the world:

> *The value that we bring to our partners is a different approach to looking and managing risk. Traditional risk factors like age and gender are static. We have a multidimensional view of risk, so we bring your health and behavior into the equation. Then we offer the interventions and share how we believe we can change your behavior to change your risk.*

One small challenge for the implementation of such a model exists: How do you learn about and quantify the behavior of your customers?

> *There's just so much more data and information today to make the way we operate more intelligent. We're a data-driven organization; it's the foundation of our entire business model, and it's key to all our products. Our founders are actuaries, so our obsession with data was there right from the beginning.*

Patient records are a data source for inferring a patient's past behavior from diagnoses and treatments. This approach may sound simple, but Surmava shares Byczkowski's point of view that deriving actionable insights from patient records is a big challenge:

> *In healthcare, tons of data is generated, but it's fragmented. Constructing a full longitudinal view of the patient's record is difficult. Even though so many systems are deployed from data collection points and intelligent devices, the problem of getting to a single version of the truth and a comprehensive view of the patient hasn't been solved anywhere in the world.*

It took Vitality a decade to actually integrate these information sources and start collecting some additional data from customers and members. Vitality helped the industry integrate with hospital systems and doctors, and they could use rich data from Discovery Health's insurance claims, said Surmava.

Negotiations were one part of the journey; convincing systems to provide data access was another. The widely adopted Fast Healthcare Interoperability Resources (FHIR) standards are used to exchange electronic health records. But when Vitality tried to integrate some of its devices, the company quickly discovered that its interfaces were not that standardized, which would require a structural solution for device classes.

Successfully negotiating agreements and solving technical difficulties were the starting points for an active engagement with Vitality's customers. Patient records show past data, but customer engagement happens in real time and is a key success factor for Vitality, so measuring and tracking engagement is an important indicator for their success. Just downloading the Vitality app doesn't count as "engagement," but if customers record that they've exercised or just taken their medications, Vitality can see that they are engaged:

> *We have successfully engaged members across diverse populations in all age groups, genders, and risk profiles. Our Vitality engagement scores are over 60% across our markets.*

Vitality used behavioral sciences to solve the engagement problem. The principle is that incentives and rewards drive engagement and behavioral change, as Surmava explained, but the company initially wondered how the level of engagement depends on the age of their customers.

Engagement can be tied to positive outcomes beyond the well-being of customers. Cost savings for healthcare are directly correlated with productivity increase for employers. One positive behavior change can trigger the next:

> *Once people start to exercise, they often start eating a healthier diet, and if they were smoking before, they often stop smoking. There's even a correlation with other behaviors like driving better or making better financial decisions. We see this effect across our different products and divisions.*

Data privacy regulations require customer consent to collect and correlate data from a variety of sources because Vitality runs a composite model. Banking contributes retail data. Healthcare adds clinical data, including data from wearables and diagnostics devices. Vitality adds behavioral data and uses advanced analytics and data mining to find meaningful correlations and trends to design and optimize their products and services.

Surmava explained how Vitality responded to COVID-19. It identified high-risk workers and ensured that infected workers received oximeters and other support like "hospital-at-home" services. Virtual consultations rapidly scaled up:

> *We already had our own platform for our members for virtual consultations. But in the space of two weeks from the first case in South Africa, we've set up the virtual consultation platform, a network of doctors, and we've opened up the platform to all of South Africa.*

Later, Vitality was a leading party to drive the vaccination initiative, working with the government to set up massive vaccination centers, which became popular across the country.

SAP provides the foundation for Discovery Health's platform, as shown in Figure 5.3, but its role goes beyond being the technology and application stack. Surmava described SAP as a partner across Discovery Health's spectrum of products and services as an insurance and healthcare provider.

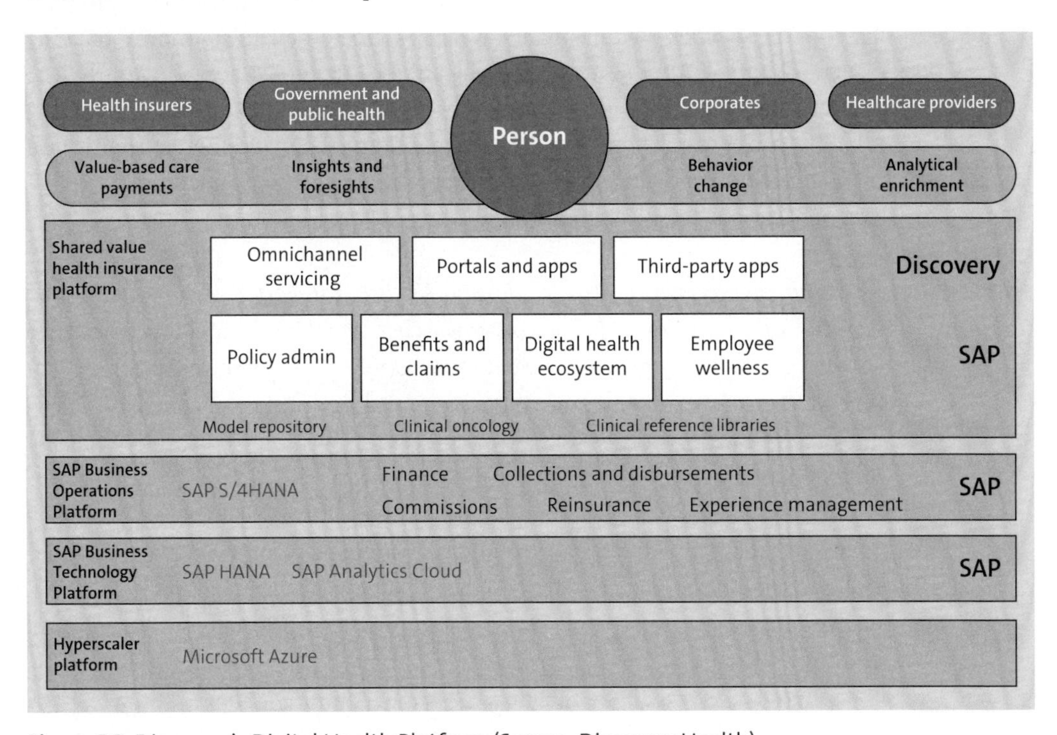

Figure 5.3 Discovery's Digital Health Platform (Source: Discovery Health)

But more importantly she sees SAP as her partner on the journey of Vitality Health International to become a software and platform provider:

> *SAP's guidance has been invaluable for me, just trying to understand the nuances of being a technology provider.*

Surmava described the history of the Discovery-SAP relationship:

> *Before our bank was launched, I would say 99% of our applications were developed in-house. In Discovery Health, for instance, I had a large development team of close to 700 people developing our applications.*

The transformation to increased usage of standard software started with Discovery Bank getting fully set up on the SAP Fioneer platform.

5.3 The Zuellig Pharma Story

Zuellig Pharma, based in Singapore, is a $13 billion distributor for most major pharmaceutical companies. Daniel Laverick is the head of digital and data solutions at Zuellig Pharma. He is driving a "digital-first organization" in a region where many of their 13 Asian markets still run immature technologies on immature infrastructures. Zuellig Pharma had to move fast when the COVID-19 pandemic changed the distribution of medical supplies and vaccines to both big pharmaceutical chains and small pharmacies. Laverick described the challenge at the start of the pandemic:

> *The pharmaceuticals industry relies on traditional, very labor-intensive processes. Sales reps would go out and visit customers. Collectors would pick up cash from the customers and take it to the bank. These were very manual processes. Overnight, there were no sales reps and there were no collections. People couldn't process orders in the offices, and so we had to pivot very quickly just to run our standard day-to-day operations.*

Zuellig Pharma, like many other organizations, had pursued a long-term plan to launch e-commerce in their heterogeneous market, but this plan quickly went into overdrive when the pandemic hit:

> *We always had a plan to launch e-commerce and digitize our order-to-cash process. We always had a plan to automate collections rather than send out collectors. But this was a 24- to 36-month plan to roll out across 13 markets. We had to hugely accelerate that and launch in eight weeks in all of the markets.*

eZRx became the Association of Southeast Asian Nations (ASEAN)'s largest business-to-business (B2B) e-commerce platform for the healthcare industry. In 2021, 11.6 million order lines were automated, with manual order lines reduced by half. Still, additional challenges awaited Zuellig Pharma. It could ship products ordered online, but the finance office's collections function was still stymied, so net working capital suffered. Once Zuellig Pharma implemented SAP S/4HANA Cloud for customer payments, customers could pay online using their standard bank protocols:

> *Finally, we had the challenge of manual, email, and WhatsApp order-taking. Without people in our offices, how would we process those orders? Enter SAP's Intelligent Robotic Process Automation [iRPA].*

The new e-commerce platform, SAP S/4HANA Cloud for customer payments, and robotic process automation for order management brought Zuellig Pharma and their customers through the pandemic. And they are set up to serve a new generation of customers—the millennials—directly. Millennials are the first generation generally comfortable with online shopping, so Zuellig Pharma is working with partners to ensure a seamless customer experience:

> *Of millennials, 70% or 80% are comfortable buying from e-commerce, online platforms. As a B2B business, our customers are pharmacy chains, hospitals, and clinics, who don't necessarily use e-commerce. But we now see the need to go B2B2C, and we're working with insurance providers and pharmacy chains to be able to go this last mile directly to the patients.*

Then, there is the issue of diversity among the 13 countries. In Singapore, electronic leaflets are starting to replace the paper leaflets in drug packages that few people read. (Those that do read the inserts can still find the information online.) This new digital approach saves on costs and is good for the environment. But other Southeast Asian markets have yet to take similar steps. Other areas in which different nations vary include drug serialization and how cold chains are operated—a necessity for many medications and, of course, for COVID-19 vaccines.

Serialization gives each individual package a unique serial number, an important measure to fight drug counterfeiting. Pharmacies and even patients can scan the barcode and get a pretty reliable green or red light that the package is genuine or counterfeited. SAP is working with a number of pharmaceutical companies on this hugely important topic.

Zuellig Pharma has a unique digital application, the eZTracker. According to Laverick, eZTracker creates end-to-end visibility along the supply chain to fight drug counterfeiting, which is a particularly common problem for healthcare in Asia:

> *In Asia, there's quite a big problem around counterfeit medicines. Anywhere between one and three out of 10 products are either counterfeit or substandard. They've been tampered with in some way. The whole idea was to build a*

product that would allow a patient to be able to scan a product via a label—be it a label that's applied or a label that's printed at the manufacturing site—and from there be able to say, 'Yes, I have a degree of confidence that it's a genuine product which has been through a secure supply chain and was stored in the right conditions.'

This blockchain-based solution is built on SAP BTP and delivers the following features:

- Product verification for patients and healthcare providers to assure key stakeholders of product authenticity and quality, facilitate patient education, and streamline product recalls
- Tracking and tracing of sales rechanneling activities to monitor gray markets, reduce revenue loss, strengthen brand trust, and comply with local regulations
- Enhancing cold chain monitoring to verify that products are stored and delivered at temperatures that maintain their quality
- Driving auto-replenishment to minimize medicine stock shortages and facilitate customer returns
- Streamlining e-product information to shift from paper leaflets to digital documentation without compromising trust

Zuellig Pharma works to improve eZTracker to stay ahead in the arms race with the counterfeiters. For Laverick, this is a continuous battle:

> *Recently we found that, for some products, counterfeiters could copy the exact packaging. The only way that it flags up is because we've gone one step further to say, 'Okay. We need to look at a digital fingerprint to compare the print quality of, for example, particular serialization codes. I can see that if this escalates further, we will have to look at other kinds of identification for more protection. Perhaps when you open a pack, there's an IoT device that sends a notice to say that this pack was opened and is being consumed.*

Laverick expanded on the cold chain challenge:

> *What we are known for at Zuellig Pharma is our vaccine distribution, and cold chain is our key strength. At the onset of the pandemic, we were probably one of the few distributors that had super cold chain, certainly for the region. Basically, we're talking minus 70-degree deep-freeze storage facilities. And so, in all Southeast Asian markets, we were basically at the forefront of the COVID vaccine deliveries. However, it was challenging to roll this out in places where you have to go quite far distances into provinces, and you don't want a lot of stock to sit in the heat with the chance of batches being spoiled.*

If you can't guarantee a robust cold chain for a long time, you have to change the game. So Zuellig Pharma minimized the time out of refrigeration (TOR) in cooperation with the vaccination sites:

> *The whole idea was to vaccinate as many people as possible in the shortest amount of time. I think it helped that we tied this in with our core SAP system and were able to allow the vaccination sites to order as and when they needed. They could track that order and know exactly when it was going to be delivered. From there, we could also very accurately plan the schedules for vaccinations.*

The pharmaceutical supply chain is still a labor-intensive operation, but when you digitize order-taking, your fulfillment operations can better keep up, as explained by Laverick:

> *Nearly 70% of our 12,000 employees are in distribution—from pickers to people driving delivery trucks. That is a huge amount of manual work, so we have been looking at automation efficiencies. Hong Kong is a great example. We have a fully automated warehouse there, run by only three or four employees. The rest are robots and drones to do stock counts. In Singapore we recently did a pilot focused on 5G deployment, augmented reality via Google Glass, and hands-free picking. With 5G, the speed at which you can look up information in SAP was huge compared to the traditional way of doing things.*

For Zuellig Pharma, Laverick saw new possibilities to apply Industry 4.0 manufacturing innovations to the pharmaceutical distribution space:

> *We've also started to look at the metaverse. How can we bring a digital twin model of our warehouse operations so that we could start optimizing and moving around certain picking areas or conduct bulk-picking versus loose-picking. What would that look like, and can we model and test it out before we actually start to do it?*

And finally there is the last leg of the distribution journey, getting the product to its final destination:

> *The final issue concerns all-around route optimization. How do we do deliveries the right way? For that, we're again looking at SAP products, for transportation management specifically, and how to optimize that part of the operation so once product leaves our warehouses, we can make sure we deliver it via the optimal route.*

As we can see, Zuellig Pharma's results from its evolving digital mindset are indeed impressive, especially given the variation in digital maturity of many of the markets they serve. But maybe it's another case of a dynamic market that offers more opportunities to push boundaries and develop next practices than what more static markets would allow.

5.4 Turbocharging Life Sciences

We've looked in depth at the downstream side of the health value chain—the distributors, the providers, and how they engage with patients. Moving up the chain, we find the manufacturers, pharmaceutical companies, and medical device companies, with their suppliers.

SAP's Mandar Paralkar talked about some of the rapid changes in life sciences, emphasizing both the need for improving value for patients and for common sense in regulations. We tried to keep pace with his rapid-fire examples:

- **From small to large molecules**

> *"Traditional pharma makes small, comparatively simple, chemically synthesized molecules. Take aspirin, as an example. Each molecule is made from only 21 atoms in $C_9H_8O_4$, and they flood your body to take your headache away and make your blood thinner. If you think that paracetamol is more sophisticated: $C_8H_9NO_2$ has one less molecule and features a little blue extra nitrogen atom.*
>
> *But the industry is moving to big, complex molecules made from many thousands of atoms; they are called biologics. Here is a random example: Adalimumab is an antirheumatic drug with the sum formula $C_{6428}H_{9912}N_{1694}O_{1987}S_{46}$. Of course, big molecules are more difficult to manage during the manufacturing process, and they are way more expensive to produce than small molecules. But they also promise great health outcomes, so patients will hope to get their (or their health insurer's) money's worth."*

- **Anti-counterfeiting measures**

> *Doctors, pharmacists, and patients need to be able to trust that they prescribe, provide, and ingest the genuine product. There are plenty of counterfeits and gray zone commerce that requires effective serialization measures. You've already discussed this with Daniel Laverick of Zuellig Pharma. But the problem extends to medical devices, and regulatory agencies are stepping up. There are investigational drug medicinal product regulations and medical device regulations for components in medical devices. If a fluid bed dryer has a certain filter*

that is clogging and causing the issue with certain part failure, agencies need to know what kind of key resources are going into the making of the drug.

- **Blurring industry boundaries**

> "*We have seen large medical device manufacturers open their ventilator design so that the automotive industry could help to scale manufacturing. Industry lines are blurring and there will be more new entrants from other industries. We are also seeing big data-related technology acquisitions like Flatiron and partnerships like 23andMe in the market."*

- **Everything from home**

> "*COVID-19 has made us an everything-from-home society, and there are even more changes in therapies, distribution of medicines, repairs, and even clinical trials. Patients can expect a technician to come to their home to service or repair a dialysis machine rather than taking that unit to the clinic. Clinical trials are looking at delivering medicines to patients' homes. Agencies have become lot more relaxed about veterinarian drugs and allowing the vets to examine pets remotely. Human clinical trials are also changing. We are seeing decentralized trials. You don't need patients to come into clinics. You can send the medicine to their home, and the data can be captured through IoT devices and then submitted to agencies like the FDA for fast-track approvals.*

Today, SAP holds a strong market position with certified enterprise management applications for the sector. For business support functions, customers are using products like SAP Ariba for procurement, SAP Fieldglass to manage the extended workforce, and SAP Concur for travel and expense management.

SAP has been accelerating new industry-specific functionality delivered as cloud extensions that keep the digital ERP core free from modifications and complex add-ons. In prior years, we have delivered the cloud solutions SAP Information Collaboration Hub for Life Sciences and SAP Advanced Track and Trace for Pharmaceuticals. Paralkar provided an update on new cloud applications around broader functionality for R&D and clinical projects, sourcing and procurement, supply chain resilience, sustainable and compliant manufacturing, and commercialization via SAP S/4HANA, focusing on three new cloud applications and one where SAP is in the early stages of developing specifications.

SAP Intelligent Clinical Supply Management gives life sciences companies end-to-end visibility into clinical supplies, from planning to production to the patient. This solution offers the following benefits:

- Higher flexibility to quickly respond to trends and new, more complex clinical trials
- Simpler, more efficient, and automated processes
- More collaboration between all clinical trial supply chain stakeholders from suppliers to life science companies to third-party logistics providers to hospitals or patients
- Reduced cycle times for clinical trial supply chain processes
- More transparency through improved site level visibility and planning

As Paralkar explained, pharmaceutical companies are looking for synergies between clinical and commercial supply chains:

> *And that is why we have built certain capabilities on SAP Business Technology Platform in intelligent clinical supply management to start with certain elements of clinical study protocol with a module called Clinical Study Master. To keep the core clean, we don't want to add new objects to SAP S/4HANA, so we designed the Clinical Study Master as a cloud application. SAP Integrated Business Planning does commercial planning. We are doing clinical trial-specific planning in this forecasting module. That helps life sciences companies decide what they need to produce based on trial enrollment rates. But then we can use the normal material requirements planning of SAP S/4HANA for manufacturing, packaging, and distribution with an add-on relevant for clinical supplies.*

SAP Batch Release Hub for Life Sciences is a cloud application that replaces today's manual legacy batch release processes that are complex, time consuming, and error prone. The batch release application is built for life sciences to meet industry-specific regulations like Annex-16[14] by ensuring that qualified personnel have easy access to relevant data and checklists to approve a batch. A lot of the data comes from SAP's supply chain and manufacturing but can include third-party manufacturing execution systems (MES) and laboratory information management systems (LIMS) calibration data. All the relevant data is brought into the repository and the related checklists to ensure that only batches that pass quality control can get the green light.

Paralkar next talked about another cloud application for Cell and Gene Therapy (CGT) orchestration scheduled to be available in 2023. CGT offers promising new options for patients, but these therapies are complex to develop, manufacture, and deliver. In contrast to traditional therapies, individualized therapies often involve extracting cells or tissue from the patient or a donor, processing the sample, manufacturing a personalized therapy, and administering the final product to the patient. If you think this is a small thing, note that material from a patient (like blood and tissue samples) and its data are becoming part of the manufacturing process.

The end-to-end process involves sending apparatus kits to hospitals to gather a patient's blood and then having the lab analyze the sample. Next comes the manufacturing process for that individual lot size and the logistics of delivering it to the patient. CGT orchestration will need to handle anonymized patient data and ensure the right patient is treated with the right medicine at the right time without error in the value chain.

Finally, Paralkar discussed how the rapid global distribution of COVID-19 vaccines showed the need for better cold chain logistics. Just as the wholesale pharmaceuticals industry has experienced, vaccine distribution is complex. Each vaccine may have a different chemical composition, but they all require a narrow temperature range in storage and in transit, all the way to the patient's arm. SAP is currently in discussion with co-innovation customers to discuss a cloud application for this type of end-to-end logistics. It will try to benefit from learnings from COVID-19 vaccine distribution especially in emerging markets without adequate refrigeration. Paralkar illustrated some of the complexities:

> *Big molecules like vaccines require temperature monitoring throughout the value chain, which often has time out of refrigeration (TOR). That needs to be tracked to ensure it is within tolerance. There is a lot of calculation done today on the shop floor with respect to maintaining the planned versus actual time out of refrigeration. IoT and connections with third-party sensors will also come into play. Not just temperature—there are additional parameters like vibration that can influence the quality (with respect to effectiveness and efficacy) of the drug along the chain. We are very early in our discussions with our co-innovation customers.*

Research and development, manufacturing and distribution processes and practices continue to evolve rapidly in life sciences. As an example, BioNTech manufactured a lot of its COVID-19 vaccine in its mRNA production plant in Marburg, Germany. Now, BioNTech is looking at modular factories housed in shipping containers for distant markets like Africa. The company described the details of its BioNTainers project in a February 2022 press release:

> *The manufacturing solution consists of one drug substance and one formulation module, each called a BioNTainer. Each module is built of six ISO sized containers (2.6m × 2.4m × 12m). This allows for mRNA vaccine production in bulk (mRNA manufacturing and formulation), while fill-and-finish will be taken over by local partners. Each BioNTainer is a clean room which BioNTech equips with state-of-the-art manufacturing solutions. Together, two modules require 800 square meters of space and offer an estimated initial capacity of*

for example up to 50 million doses of the Pfizer-BioNTech COVID-19 vaccine each year. The BioNTainer will be equipped to manufacture a range of mRNA-based vaccines targeted to the needs of the African Union member states—for example, the Pfizer-BioNTech COVID-19 vaccine and BioNTech's investigational malaria and tuberculosis vaccines—if they are successfully developed, approved, and authorized by regulatory authorities.[15]

Some experience and exposure to the usual business processes and practices of the pharmaceutical industry, their suppliers, and regulators are required to appreciate the fundamental changes in pharmaceutical development, manufacturing, and distribution chain. SAP will continue to challenge (and be challenged to drive) innovation in this rapidly evolving sector.

5.5 The Operating Room of the Future

When we discuss the Resilient Supply Network megatrend in Chapter 9, we describe how Industry 4.0 thinking has influenced the industrial shop floor and the warehouse through robotics, IoT devices, and conveyor belts working in harmony to make human workers far more productive. Similarly, these technologies are bringing new productivity advances to operating rooms for surgeons, specialists, and nurses.

The department of urology at UKHD initiated the OP 4.1 project to develop a prototype digital platform for the operating room of the future that could better put technologies to work, such as robotics and IoT devices. Cooperating partners on this project include the German cancer research center DKFZ, the endoscope maker KARL STORZ, mbits imaging, SAP, and Siemens Healthineers. The project is funded by the BMWi, Germany's Federal Ministry for Economic Affairs and Energy.

Byczkowski of SAP added some perspectives on the project:

> *One of the goals of the project is to show that we can simplify the market introduction of software-based innovations into clinical practice. For most software innovations, the barriers to market entry are currently still too high, which is why many do not make it into the operating room and into the daily clinical routine. This is often due to a lack of access to relevant data from medical devices or due to limited opportunities for professionally implemented sales and support.*

The OP 4.1 platform can become a hub for exchanging data between hospitals, software developers, and device manufacturers. A paper in *JMIR Medical Informatics* described the progress of this project:[16]

> *We demonstrate the successful development of the prototype of a user-centric, open, and extensible platform for the intelligent support of processes starting with the operating room. By connecting heterogeneous data sources and medical devices from different manufacturers and making them accessible for software developers and medical users, the cloud-based platform OP 4.1 enables the augmentation of medical devices and procedures through software-based solutions. The platform also allows for the demand-oriented billing of apps and medical devices, thus permitting software-based solutions to fast-track their economic development and become commercially successful.*

Byczkowski noted that this prototype is applicable far beyond UKHD, and he expected that this business model will eventually serve the wider public:

> *This should enable a business model for a multisided market that offers billing options and thus real commercial value-add. We are introducing various device manufacturers, clinics, and independent software developers to each other so that they can exchange services and try out various business models. Such a platform would create even more data, which can be used to improve patient treatments further, and to further enrich patient-centric data networks.*

The co-innovation project also led to the creation of four starter apps created by consortium partners to showcase to power of the platform created by SAP:

- Context-sensitive augmented reality (by DKFZ): The app assists surgery by displaying target or risk structures, like tumors, in real time.
- Live perfusion measurement (by DKFZ): The app evaluates the image data of different spectra and visualizes the flow of blood, for instance, in blood vessels underneath the tissue or in the blood supply going to the organs.
- Precision puncture (by DKFZ): Computer-aided information on the position of the puncture needle, combined with modern imaging, enables precise and navigated tissue punctures.
- Mobile information service (by mbits imaging): Transfer of current information about the surgical progress from the platform to mobile devices, adapted to the needs of the recipient.

5.6 Patient-Centric Data Network

Byczkowski has a bird's-eye view of all the players that influence healthcare and their interactions. He envisioned a network for life sciences, healthcare, and beyond to inspire innovative business models, products, and services, all eventually supporting processes across industry boundaries, as shown in Figure 5.4.

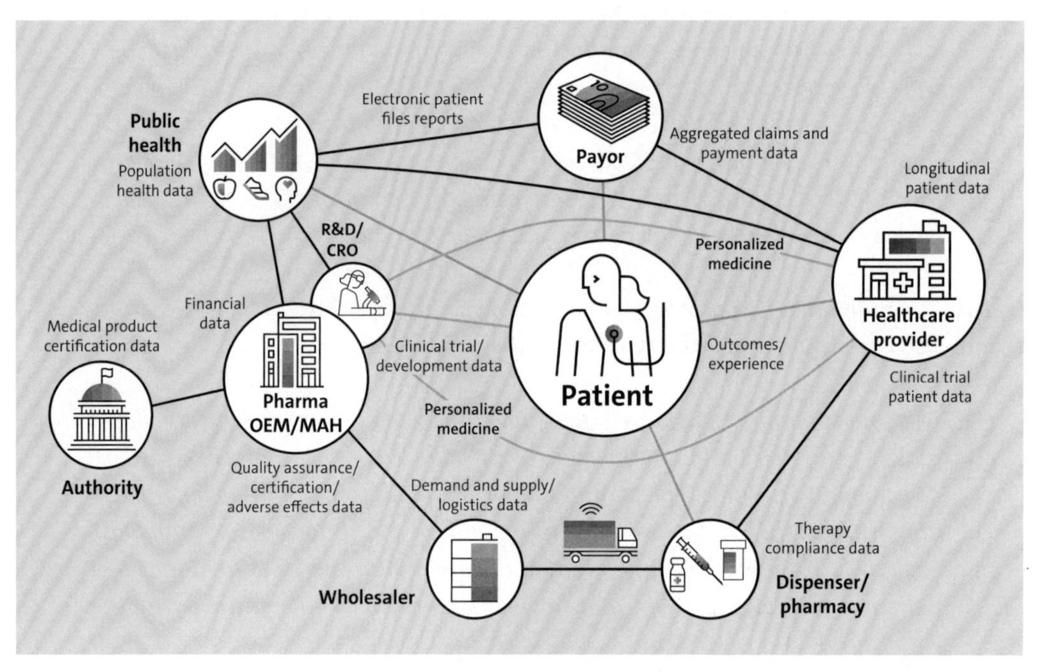

Figure 5.4 Patient-Centric Data Network (Source: SAP)

But Byczkowski acknowledged that we are a long way from a harmonious network thanks to a number of competing (or at least incompatible) priorities and motivations:

- Healthcare providers are dedicated to serving patient needs, but patients are neither "customers" (since their health insurance provider pays on their behalf) nor "consumers" (since they do not choose to consume the products and services and thus there is little potential for upselling or cross-selling).
- The entire ecosystem depends on indirect connections, which creates a lot of friction and consumes a huge amount of time and resources. Multiple stakeholders within a highly regulated environment with sometimes conflicting priorities may make reaching the declared common goal—helping patients get healthier—difficult.
- Patient data/real-world evidence is not equally distributed nor is it seamlessly accessible due to both privacy and other regulations, as well as asymmetric interests and investments. For example, hospitals and research organizations spend large amounts of resources on cleaning, annotating, and storing data but may suffer from a lack of structured data and semantics.
- New business models may be easy to envision but hard to implement due to the many interwoven dependencies and interests. But there is a clear, significant value

to be achieved by a more connected and more data-driven health ecosystem and with digital innovations and new technologies, and the key enablers are available today.

Next, Byczkowski proposed some guidelines for key stakeholders shaping this network:

- A data- and value-driven approach will establish connections on the basis of standardized data entities (such as "patient," "treatment," "drug," "trial," and "payor"). The digital end-to-end processes will follow the data.
- All stakeholders in this ecosystem will treat their data as proprietary assets until commercial models incentivize sharing data and information.
- A digital network with clear data access and usage rules will harmonize, enrich, exchange, and commercialize data in a safe, secure, and trusted environment and in a commercially viable and compliant manner between individual contenders.

This approach could lead to better information for planning clinical trials; easier estimation of market sizes for new drugs; and more successful training of machine learning algorithms for diagnosis, treatment decisions, treatments, and follow-up. These activities would enable more successful implementations of new business models like pay per use, pay by outcome, and, eventually, everything as a service.

Byczkowski believes SAP is best positioned in the industry to drive such an approach due to its neutrality as well as the existing broad and deep customer relationships across all involved industries. As we have seen, SAP has relationships in very diverse health systems and across continents. Byczkowski established three dimensions to consider when moving to such a data-driven economy for better health outcomes:

> *The first dimension is function. Focus on function and identify specific use cases within an enterprise. SAP offers functional solutions like SAP SuccessFactors or SAP Fieldglass that have a well-defined scope.*
>
> *The second dimension is process—in other words, defining, implementing, and optimizing well-defined processes within and beyond the four walls. Here the supply and demand insights and actions that you focus on are not only from within the company, but also from adjacent companies, and perhaps even from companies that are some steps removed in the value chain. This is where SAP's value proposition of the intelligent enterprise comes into play.*
>
> *The last dimension is data. We think about data exchange across networks. We look beyond specific well-defined processes and look at data wherever it is available and how it can be used to create new business models, innovative products, or trailblazing services elsewhere. These new business models might end up creating completely new processes across industry boundaries. Here's SAP network thinking by industry is connecting those intelligent enterprises.*

Byczkowski was enthusiastic about the success of large platform providers in other sectors:

> *When you look at Amazon, Alibaba, or Apple, you realize that the platform business model offers great commercial opportunities with a global reach. Different vendors and types of vendors can be all brought together on one single platform offering excellent consumer experiences.*

5.7 Lifelong Health with SAP

Life expectancy at birth has never been higher. It ranges from 54 years in the Central African Republic to 85 years in Hong Kong, with a global average of 73 years.[17] This is good news for us individually but puts a big strain on our healthcare system: In the US, almost one-fifth of the GDP went into the health sector in 2020.[18] A high life expectancy doesn't automatically come with great health, and the leading causes for death in high-income countries correlate to lifestyle. The World Health Organization frames it candidly:[19]

> *It is important to know why people die to improve how people live. Measuring how many people die each year helps to assess the effectiveness of our health systems and direct resources to where they are needed most. For example, mortality data can help focus activities and resource allocation among sectors such as transportation, food and agriculture, and the environment as well as health.*

This insight is influencing the lifestyle of millions. For example, the "quantified self" movement is creating a growing market for all types of health monitoring devices, exercise regimes, and nutrition products, and advice.

5.7.1 Mental Health and Resilience of the Healthcare Workforce

Providers seek to deploy the right skills to service demands, reduce retention risk, support employee development, and monitor their experiences and stresses while increasing compliance to standards, enhancing workforce engagement and performance, and improving job satisfaction and morale.

The following solutions that contribute to healthcare workforce mental health and resilience might be part of your SAP-run IT landscape:

- SAP SuccessFactors solutions support activities for managing competencies, skills, and best practices and contains all the necessary talent management modules.
- SAP Fieldglass manages contingent workers and worker profiles.

- SAP Business Technology Platform is a platform for delivering dynamic web applications and for integrations.
- SAP Analytics Cloud enables the aggregation and visualization of data from other modules of the solution.
- Qualtrics XM helps you capture employee experience survey data.
- Safe Workspace Management by SAP partner DXC Technology addresses the increasing business needs of the healthcare industry by providing tools to manage, support, and care for healthcare workers.

5.7.2 Innovative Software-Based Medical Solutions

This focus area addresses the need to quickly and efficiently translate software-based medical innovations into clinical routines and to optimize patient treatment through smart-assisted procedures via a single platform.

The following solutions that contribute to software-based medical innovation might be part of your SAP-run IT landscape:

- SAP Business Technology Platform, Cloud Foundry environment, is the backbone for the full solution development, including data management and analytics.
- SAP Subscription Billing provides the basis for billing scenarios, such as pay-per-use and invoicing, for medical device usage and application usage.
- SAP Conversational AI works with the speech-to-text service to the support voice recognition-based control of the platform's dashboard.
- SAP API Management technology is part of the open SAP architecture and provides standard technical communication for innovative solutions built by SAP partners and customers to integrate with SAP's Intelligent Suite.

5.7.3 Supply Chain Transparency for Pharmaceuticals and Remedies

This focus area addresses the need for an all-in-one digital drug and vaccine supply distribution and administration solution to simulate, plan, and manage the journey of products from regional distribution centers to vaccination centers and, ultimately, into people's arms.

The following solutions that contribute to supply chain transparency might be part of your SAP-run IT landscape:

- SAP Integrated Business Planning for Supply Chain synchronizes supply chain planning in real time, from sales and operations planning to inventory and supply planning, and ensures that the right pharmaceuticals are available in the right location at the right time.

- SAP S/4HANA Cloud unifies and enhances ERP functions, including procurement, inventory management, supply chain, and logistics to handle pharmaceuticals in a compliant way.
- SAP Analytics Cloud delivers business intelligence, augmented analytics, predictive analytics, and enterprise planning in a single, cloud-based solution to optimize the supply of pharmaceuticals.

5.7.4 Inventory and Bed Management

Medical staff have always needed a way to inventory ICU beds on the fly so that emergency room patients can receive life-saving treatment quickly and efficiently.

The following solutions that contribute to inventory and bed management might be part of your SAP-run IT landscape:

- SAP Business Technology Platform forms the backbone of the solution and keeps it flexible and scalable, while aggregating data, managing role-based access, and delivering input to the analytics functionality.
- SAP Analytics Cloud can manage the conversion of aggregated data into actionable insights, for instance, in a breakdown of ward capacity combined with geolocation data and a visualization of occupancy rates over time.
- The SAP Fiori user experience (UX) displays user options and analytics in an attractive and accessible format on all devices, creating a user experience that allows healthcare personnel to work with patients, and not with computer screens.
- SAP Cloud Application Programming Model is the main framework of the project and provides the main functionalities while connecting to other SAP services used in the project.

5.7.5 Value-Based Care

Value-based care links financial reimbursements to outcomes achieved during an episode of care with a patient. In value-dependent pricing, risk-bearing entities must collaborate more closely and take a more active role in optimizing patient care. Incentives to engage patients in preventive measures or to enhance the holistic care experience require building more direct relationships between the entities involved in the delivery of care to the patient.

The following solutions that contribute to value-based care might be part of your SAP-run IT landscape:

- SAP Experience Management facilitates the capture of patient-reported outcomes and patient-reported experiences, which are necessary data points to support value-based care and contracting, while complementing operational and clinical data points.

- SAP Patient Management provides information to support patient-level cost, billing, and statutory data points required to support value-based reimbursement: Features include patient demographic data capture and updates, insurance relationship data capture and updates, admissions and discharges, transfer synchronization, services recording, creating basic documentation for diagnoses and procedures, reconciliation, and billing status.
- SAP S/4HANA inventory management connects to clinical information systems to facilitate cost accounting and reporting and support accounting on material cost levels that is required for reimbursements.
- SAP Ariba facilitates the acquisition of medical supplies, consumables, and specialist equipment. Sourcing new suppliers is often overly complicated due to strict regulatory requirements, so having a list of vetted and approved medical suppliers speeds up processing, increases access, and reduces costs.
- SAP Analytics Cloud's Digital Boardroom provides risk-bearing entities with a unified view of operations against their service level agreements (SLAs).

5.7.6 Business Process Efficiency

At the beginning of COVID-19, orders for medical supplies and later vaccines exploded. One challenge providers and healthcare suppliers faced was the reduced manpower to quickly process, fulfill, and deliver a very large number of orders coming in.

The following solutions that contribute to business process efficiency might be part of your SAP-run IT landscape:

- Robotic process automation (RPA) implements automated workflows to reduce the time between incoming order and order fulfillment.
- SAP S/4HANA is the backbone for efficient end-to-end processes, from order entry and logistics to fulfillment and financial settlement.
- Blockchain technologies provide innovative and secure ways to do business, for instance, implementing the Safe Order app so health procurement experts can quickly put in their orders.

Chapter 6

The Future of Capital and Risk

Of Bear Hugs and Sushi Bonds

In the 2015 movie *The Big Short*, actors Margot Robbie and Selena Gomez and celebrity chef Anthony Bourdain explained complex financial terms like synthetic collateralized debt obligations (CDOs) using simple, everyday examples.[1] You may laugh at the over-simplification of arcane economic lore, but consider that Hollywood has come a long way since *Mary Poppins*, set a century earlier, which showed two young kids causing a run on a London bank over a twopence.

In reality, the average person understands very little of the complexity of financial markets. Even industry insiders struggle to keep up with terms unique to financial markets, like "bear hugs," "sushi bonds," "suicide pills," and the "dead cat bounce."[2]

Financial institutions range from local savings and loans institutions to global reinsurance companies. The global financial services sector was estimated at $26 trillion in 2022.[3] Just in the last few years, we have seen even more fragmentation in financial institutions and products. As Gartner predicted back in 2018,

> *By 2030, 80 percent of heritage financial services firms will go out of business, become commoditized or exist only formally but not competing effectively.... These firms will struggle for relevance as global digital platforms, fintech companies, and other nontraditional players gain greater market share, using technology to change the economics and business models of the industry.*[4]

Similarly, in June 2022, the Bank of England published its first assessment of the preparations of eight major UK banks to comply with the solvency and reserve requirements of the Resolvability Assessment Framework (RAF). The assessment claimed that none of the large UK banks was "too big to fail" and tried to prepare the public for yet more change in the global landscape of financial services.[5]

Conversely, a *Barron's* guest columnist in 2021 asked, "One wonders how we'll look back on these market developments a decade from now. Will SPACs, cryptoassets, and mobile trading apps (like Robinhood) be seen as hybrids that emerged in the antechamber we are living in now?"[6]

This chapter focuses on the banking and financial services space, where the stakes are always high and digital innovation percolates throughout the sector. Banking goes back in time to the dawn of history in Mesopotamia with the first coins and evolved into its modern form in Renaissance Italy in the 14th and 15th centuries. Today, fintech startups attack the foundations of a banking system that is rooted deep in history and epitomized by marble-clad buildings. The insurance industry also looks back on thousands of years of history, and the modern concept of property insurance emerged after the Great Fire of London in 1666.

This deep history is not surprising if we take a step back and look at the fundamental role of financial services in most economies: Virtually every business transaction combines a transfer of products or services in exchange for monetary compensation, and this transfer frequently comes with a transfer of a risk that can be quantified in terms of probability and impact. However, innovative fintech and insurtech players are challenging hundreds of years of business practices and use digital technologies to disrupt these pillars of our global economy.

SAP is largely known as an ERP vendor in industrial markets, but we are also an influential player covering a surprising set of spaces and niches in the vital and fluid financial services sector:

- More than 14,000 banks in 150 countries are SAP customers.
- Half of the world's central banks and 40% of stock exchanges run on SAP.
- SAP banking systems process more than 1 billion financial transactions every day.
- More than 1,000 banks and non-banks run SAP's core banking solutions, and more than 1,250 banks run SAP solutions for financial accounting, controlling, and risk management.

6.1 Convergence and Crosscurrents

Stuart Grant, head of capital markets at SAP, is part of the team led by Falk Rieker, global head for the banking industry business unit at SAP. Grant offered his perspective on the sector's shifting roles and the larger issues of regulation and transparency—issues that younger cohorts are especially interested in—and what he sees as the industry's opportunities for wider leadership:

> *Let's face it—in recent years, capital markets haven't necessarily fulfilled the role that they traditionally have. In 2020, about 250 fintech startups managed to raise $50 billion in funding outside of the traditional capital markets environment. In North America and in Europe, more companies are choosing to stay private rather than go public because of the burden of reporting and transparency. Ironically, in some respects, this is going the wrong way from a societal point of view. The generations that are coming into the workforce now and are bringing the new wealth into the financial market's environment*

want to see more transparency, more choice, and more self-service capabilities. There's a lot that financial services can learn from other industries—like retail, for example. But at the same time, the industry has the potential to be the catalyst for change, and to support a lot of the other things that are reshaping most other industries.

Rieker has outlined three main trends across the financial services sector—industry convergence, competition for talent, and business networks—that impact industry-wide priorities of hyper-personalization, intelligent operations, cryptobanking, performance, and compliance.

As Rieker commented, "Traditional industry boundaries are blurring. The financial services industry is expanding into new areas while also consolidating in others." For example, when Goldman Sachs finally entered consumer banking in 2016, it heralded the move with an announcement: "As one of the few large banks without a legacy consumer business or infrastructure, Goldman Sachs is uniquely positioned to redefine how financial services are distributed and consumed."[7] It has proceeded to partner with Apple on its innovative Pay credit card and its "buy now, pay later" installment service—new moves that fly in the face of Goldman's long-established identity.

Rieker also noted that the dramatically named "war for talent" is being experienced across industries, beyond the limits of the financial services sector. Not only are labor shortages resulting from the COVID-19 pandemic, but organizations in this space are struggling to attract both diverse and young talent. In addition, one implication of the increased popularity of hybrid work models is decreased demand for commercial real estate, in which insurance and banking companies have traditionally been heavily invested.

These conditions have put even financial sector giants in a tight spot. "Not only do they face the threat of attack from fintechs," said Grant, "but they are also unable to operate their legacy business without the right staffing."

Finally, our financial services customers closely watch the emerging industry networks to develop and capture new business opportunities. Beyond its business network portfolio of the SAP Business Network, SAP Concur, and SAP Fieldglass, among others, we are squarely in the middle of conversations about industry-specific networks like Catena-X across sectors. Customers in the financial sector want to discuss how SAP is shaping ecosystems, how we partner, and which lessons we have learned.

Anton Tomic, who covers the insurance sector and is based in Düsseldorf, Germany, talked about new industry networks:

> *Our insurance customers are looking at mobility of the future, health of the future, housing of the future, travel of the future, and wealth management of*

the future. They are embedding themselves and looking to increase their services, expanding beyond their traditional insurance premium revenues.

Grant drew on an example of how a toy company expanded into a completely different industry:

> *It's a bit like when Lego realized their competition wasn't other toy manufacturers, but rather the likes of Netflix and Amazon and Disney because what they were actually competing for was the time and attention of their customer, who is often a child. And so that's why they started creating Lego movies.*
>
> *We see the trend that financial services get distributed and commoditized by other industries, disintermediating banks, and insurers from their customers. Can that trend be reversed? Can financial services organizations actually bring capabilities from other industries into their portfolio and differentiate that way?*

Meanwhile, the financial services industries have established industry-specific priorities. Rieker defines one of them, *hyper-personalization*, as individualized customer engagement, citing the use of robo-advisors, intelligent call centers, sales guidance, and digital twins in the insurance industry.

Grant noted opportunities around behavioral banking, which uses data to help consumers set financial goals and then rewards them for positive steps:

> *Personalization also extends to customer contracts and fee structures. Financial service providers need to better align with the commercial terms that they have with their customers. Discovery Bank calls their model 'behavioral banking,' and it is all about personalization and rewarding customers for responsible and healthy behaviors.[8] Or take our SAP BRIM product for billing and revenue innovation management that is used by many of our investment firm customers. It enables them to roll back from the general ledger environment to accounts receivables and billings, the revenue engine. It helps them plug their revenue leaks.*

Rieker connected improvements in intelligent operations with cloud computing and process optimization:

> *We see banks and insurance companies increasingly investing in optimizing their operations. That is very much aligned with a move to the cloud. When you make that move, you also look at process optimization. Our SAP Business Technology Platform with machine-learning capabilities plays a role here.*

Many financial services customers pursue a private cloud strategy, and there are others running a hybrid on-premise, private cloud, and public cloud environment. We believe in giving customers the options that make sense for their business and IT strategy, and hybrid cloud is currently the dominant model:

> *With our RISE with SAP initiative, we can support all the cloud options. It's about efficiency, standardization, and automation. That's what we focus on: bringing our customers as fast as possible to the cloud. Otherwise, they won't be able to keep up with their competitors in the long run.*

Combined software and services offerings from SAP such as RISE with SAP are facilitating financial service industry customers' investment in intelligent operations. Tomic cited three examples of SAP customers in the insurance industry using RISE with SAP for their business transformation:

> *A German insurance company is transforming the office of the CFO using RISE with SAP. They are not just replacing the technology; they are defining a new model for the finance function. A Swiss financial services firm is taking a different approach. They are using RISE with SAP just to turn their CapEx into an OpEx model by moving to a hyperscaler.*

The flexibility of the RISE with SAP model isn't limited to insurance giants or even to customers with contemporary ERP systems. Tomic noted:

> *A new SAP customer is using RISE with SAP in yet another way: They are an insurer with about $2 billion in premiums. They have 30-to-40-year-old back-office applications. RISE with SAP gives them the flexibility to handle the growth they expect.*

Cryptobanking, or the management of digital currency assets,[9] is another priority for SAP's customers in the financial services space. According to Rieker, SAP helps with the accounting side and the custodial side of managing cryptocurrency accounts.[10] You may have heard that it doesn't take a banking crash to lose cryptocurrencies; a smartphone crash can be enough. Rieker explained how VAST Bank protects cryptocurrency holders:

> *That information is then stored by the bank, and this has significant advantages because if you lose the key at Coinbase then it's gone. Here, you don't lose your assets with the key because the bank has the key as well.*

While much of Rieker's focus deals with the "crypto," "digital," and "virtual" aspects of financial services, Anja Strothkämper, vice president of SAP Agribusiness and SAP Commodity Management, enables her customers to buy and sell physical farm products. She differentiated between exchange-traded and non-exchange traded products:

> *When it's exchange-traded like grain or corn or oil seeds, the price is typically linked to the exchange, and that's where my commodity hat comes in. SAP S/4HANA solutions cater to these index-based prices, so we model all the price curves, price quotes, cash quotes, futures curves, and we have built integrated risk management.*

She pointed out the criticality of ensuring accurate risk calculations, which demands newer IT systems. The war in Ukraine has definitely raised sensitivity and awareness among many executives:

> *If you're a CFO of a consumer goods company or a trading company, and you cannot prove to your shareholders or financial analysts that you have your market price risk under control, I promise you your interest rates will go up and your credit rating will go down. And try to explain this to your board and to your investors.*

She emphasized that consumer product companies on the processing and manufacturing side are in dire need of this functionality. But she also found that companies in the trading business who run old systems need to take action to keep their commodity risk exposures under control.

6.2 Banking: A Spectrum of Customers

When you talk to Rieker and his team, you hear mention of clearing banks; new insurance models for emerging e-mobility markets; and ecosystems around real estate, healthcare, and retirement assets. You'll also hear about SAP's core banking and insurance products, which are now managed by SAP Fioneer, a joint venture with Dediq GmbH.

Rieker can rattle off the names of customers that are using the SAP Fioneer functionality:

> *Core banking, in our definition, is lending, deposits, checking accounts, and payments. We have hundreds of customers, and they range from large customers like the Commonwealth Bank of Australia and Standard Bank of South Africa, to very small, specialized organizations. We see a large opportunity*

with fintechs and with small to mid-sized banks to implement our solutions in the cloud. Replacing core banking applications in larger organizations is more complex, quite similar to the slow migration of core ERP functionality in many industrial markets.

Rieker described a significant shift in the banking sector. Larger banks are moving away from servicing retail customers with compelling solutions and regard them primarily as a source of credit card charges and overdraft fees. As a result, he said, newer, more agile banks like Discovery Bank occupy a certain market segment and position themselves as financial coaches for customers:

> *Their mindset is, 'How do we increase the number and quality of touchpoints with our retail customers?' Discovery Bank offers dynamic interest rates, they come up with individualized offerings based on customer status and combine their services with a very attractive loyalty program.*
>
> *Their banking customers experience differentiated services. Discovery Bank aims to fully onboard new clients in five minutes, conduct instant funds transfers, and provide next-day credit card delivery, among other benefits. Their insurance customers see a similarly different level of service. Where traditional insurance involves simply paying clients when they file a claim, Discovery focuses more on the preventative aspect of healthcare in line with a growing and aging population.*

Grant noted that, by offering financial coaching, banks like Discovery Bank attract customers who are seeking that type of service—ultimately reducing the risk profile on its books. Grant added:

> *Discovery attracts customers who want to be engaged in the type of services that they offer. As a financial institution, they are far less of a systemic risk than traditional banks because of the nature of their loyal client base.*

A number of SAP banking customers have settled into unique niches. Rieker cited Compartamos Banco, a Mexican bank that has been a pioneer in extending microloans to women across Latin America:

> *They offer group loans, typically for a group of 10 women with microbusinesses. On average, they are of four-months' duration and for around $200. But if one of the women can't pay an installment, the others need to stand in for her. These loans are unsecured, yet amazingly they have a default rate of less than 2%.*

Rabobank, the large Dutch bank, started issuing Renewable Energy Green Bonds in 2016 and has since introduced a complete Sustainable Funding Framework under which several types of sustainable financial instruments can be issued (e.g., bonds, loans, derivatives, commercial paper, certificates of deposit, etc.).[11]

According to *Forbes*, Vast was the "first federally chartered bank [in the US] that allows people to buy, sell and exchange crypto directly from a bank account."[12] Because crypto prices require much more granular fraction values than traditional currencies, the decimalization feature in SAP system is particularly attractive to banks like Vast, Grant said:

> ***You've got a much higher level of fraction base currency values, which the SAP systems can handle out of the box, and so that enabled them to be able to add a cryptocurrency account and obviously support that customer.***

Supporting eight decimal places for a currency fraction may not sound like a big deal, but legacy banking systems can't do it. Since this capability is deeply embedded in the system foundation, you would have to rebuild huge chunks of your banking applications—just for a (currently) small niche of the business.

Meanwhile, the arrival of increasingly powerful analytics functionality is supporting banking customers such as Lloyd's Bank in the UK. Grant explains how the Union Bank of Philippines uses analytics:

> ***They use data intelligence to test the reliability and suitability of their credit risk models because they want to be the bank in the region that is known for protecting their customers. And the only way they can do that is to ensure that when they onboard customers, they've got accurate credit risk modeling.***

6.3 The Bank of London Story

We continue to explore fascinating niches in the financial industry as we turn to discuss the Bank of London—which Grant called "the first new UK clearing bank in over a century."

The Bank of London was launched in November 2021 and boldly calls itself a "leading-edge technology company." Founder and CEO Anthony Watson introduced himself as "primarily a technologist, not a banker." He started his career at Microsoft, ran international technology for Wells Fargo, and became the chief information office (CIO) for Europe for Barclays Bank and later for Nike.

For Watson, technology is just an enabler, and the Bank of London has ambitions to shake up financial clearing by modernizing what he called "a critical, arcane, and not-well-understood cornerstone of the global financial system." Watson explained that any movement of money within the global financial ecosystem requires a clearing bank to do two things: to clear (i.e., to process, move, and net obligations from the movement of funds from one institution to another and via central banks) and to settle (i.e., to disburse the irrevocable obligations resulting from fund transfers across institutions and via central banks). Every bank, payment company, and corporation globally must use a clearing bank, directly or indirectly.

Watson's assessment of the current inefficiencies of the clearing house system was blunt:

> *Say I was to send you 50 pounds today; you'd likely get the money within 24 hours. But your bank may not settle with its clearing bank for up to 40 days. They don't know what the position is in real time. If you look at the Western world, 30% of payments are delayed by up to 60 days.*
>
> *It's bonkers. We're talking about a massive amount of money here, basically $2.5 trillion annually. Seventy-two banks (four in the UK, 24 in the EU and 44 in the US) control the movement of $2.5 trillion. It's a very high margin business mainly driven by the fact that banks are massively manual still with armies of people—and their technology is generally antiquated, expensive, and at the end of life.*

Watson isn't alone in his conclusions that a new kind of bank can make a monumental impact in the industry. Harvey Schwartz spent his career at Goldman Sachs and is group chairman and non-executive director at the Bank of London. He commented:

> *During the great financial crisis of 2008/2009, I saw first-hand how the legacy payments, clearing and settlement processes that are at the heart of the global financial system contributed to bringing the world's economies to their knees through their inefficiencies and inherent liquidity risk. Fundamentally, banking is an immensely complex data problem. The Bank of London is the solution.*[13]

Watson spent four years studying the market, applying for licenses and certificates, and developing a resilient and blindingly fast architecture with nine patents to back the company's efforts. To develop this concept, Watson and the Bank of London have partnered with SAP to design its real-time software architecture, meaning that all transactions occur in real time without periodic batch processing:

> *We built everything to be straight-through processing, so when we launch in the UK, we'll be the fastest payments partner for any institution in the UK—by far. We've independently tested our core payment transactions. We have the fastest solution in the marketplace.*

After years of trying, Watson finally got the right to join the exclusive club of clearing banks, and he has chosen a narrow specialist path:

> *We are certainly in British history—but we believe also in the Western world—the first bank that has a non-leveraged balance sheet, meaning our clients' money is not lent out and we keep it 100% safe at the Bank of England.*

Beyond speed, the narrow focus may also give the Bank of London a significant cost advantage. Although "heritage" banks designed to make loans can benefit from the highest interest rates in years, they also require large employee populations to service them. Watson explained that his organization, focused on clearing only, can be highly agile and very competitive:

> *The other clearing banks require much more capital on their balance sheet to service their loan book. Because we don't make loans, our capital costs are minimal compared to a heritage bank. Plus, most of the major players are still on mainframes. They have branch real estate costs. On a transaction basis, it looks like I'm taking less of a cut, but I'm actually far more profitable than those incumbents. It means I can go into the market at a price point they just can't compete at.*

Watson said this focus will help him in two ways: differentiating himself clearly from incumbents and, just as important, reminding him that his mission is to redefine clearing, not become dependent on the incumbents. He said that many fintech startups fall into this trap:

> *Payments companies set out to redefine banking. Instead, they become another intermediary front end, bogged down by the legacy incumbent they are shackled to. Many fintechs became wholly dependent on them to supply the most basic of banking services, such as remittance, accepting deposits, and operating accounts. A third of them have faced regulatory action over the failures of their current banking partners, yet 44% of them stay with them because switching looks painful. They are beholden to these incumbents, along with their compliance, technology, people, and the inherent risk they carry with their weak tier-one capital structure.*

Discussing the Bank of London's technology decisions and SAP's role, Watson said the first time he was highly impressed with SAP software was at Nike, where 10,000 SKUs per season needed to be processed. He realized its applicability to financial services then, and years later when he was at the Bank of London and had winnowed 17 system choices down to 3, he was sure SAP would make the cut—and we did. Watson said that, while numerous new banking solutions were popping up all the time, when one of the Bank of London's customers is handling massive volumes, it needs an established, tried, and tested powerhouse solution:

> *On a low transaction, low-cost basis, many new solutions probably work well. But if you need something battle-hardened, battle-tested that has significant volumes... Think about Lloyds Bank, which is one of the biggest clearing banks in the UK. Twenty percent of the UK's GDP probably goes through it. It's not going through a cloud-based startup bank. It's going through an iron-clad, hyper-scalable processing engine because Lloyds can't afford to get payments wrong. Ever.*

Watson said that SAP's ability to see the big picture in banking has made his clearinghouse the first he knows of to run SAP solutions from end to end:

> *We default to SAP products. If there are two in the market and one is SAP, we default to it because, from a full integration, straight-through processing perspective, the benefits even outweigh some limitations. Straight-through processing is the gold standard now in financial services, and not many banks are capable of that.*

Watson then described his high-level architecture, as shown in Figure 6.1:

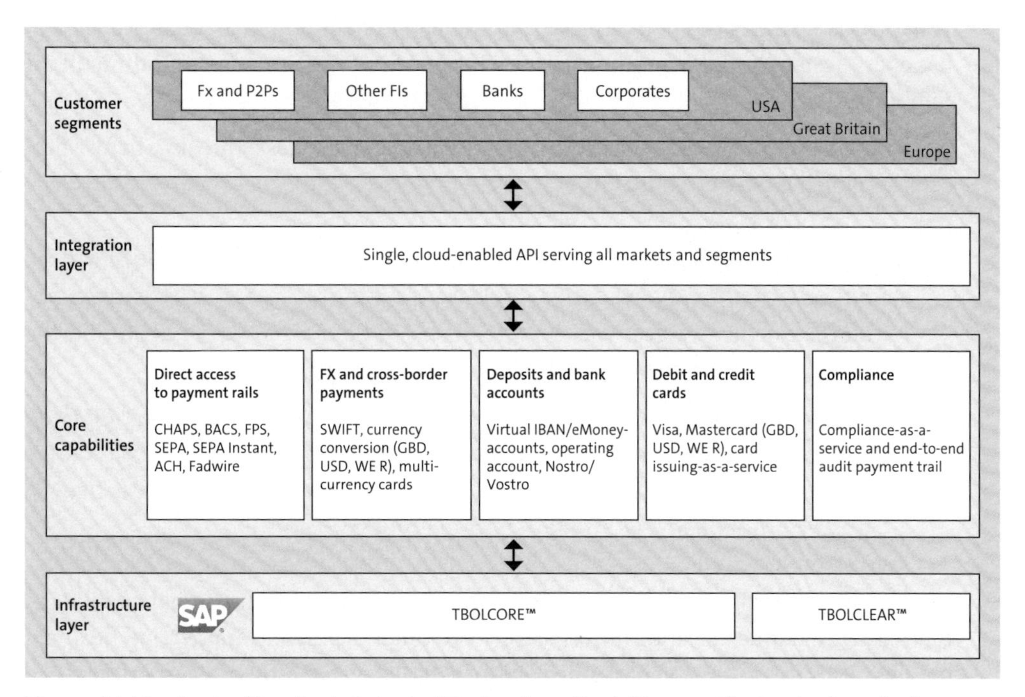

Figure 6.1 The Bank of London's Patented Technology Stack (Source: The Bank of London)

> *We spent two years making sure we had straight-through processing for our clients; that hasn't been easy. But the really hard stuff to get right is the hyper-scalability and the hyper-transaction volume, because no one processes transactions quite like financial services. We have to be always on and we can never fail to process a transaction. It's 100%; 'five nines' isn't good enough in this business.*
>
> *TBOLCore is our patented core, and we also hold a patent for our new clearing product, TBOLClear, which is game changing for the industry.*

He said the Bank of London is distinguished in its field by its straight-through processing:

> *TBOLClear runs in real time because we have no balance sheet issues. We move money in real time, not like other large banks. We also send the compliance and multiple other data sets with the transaction in real time, with the individual transaction. The receiving institution can immediately see whether the receiver is a 'bad guy' and respond accordingly. In the current model, they would pay out the bad guys, and only find out months later that something was wrong. When detached compliance information trickles in, they are scrambling to contain the damage—and on top, they may, at the very least, get fined by the regulator.*

Watson has heard in the marketplace that the Bank of London is the only player that has every system built for straight-through processing, but of course this is difficult for him to verify.

Watson said the core banking platform is what ultimately makes or breaks a bank's processing, which is why he chose SAP:

> *SAP has always been known in the marketplace as a super-reliable platform, but one that is very expensive, and very difficult to change. We've taken out all the unwieldy pieces we don't need and put our cloud wrapper around the core processor, which we kept intact to become hyper-agile.*
>
> *SAP has its limitations, but nothing is as stable or scalable as its core banking platform, at least that we've seen in the marketplace. It is unwieldy, and we spent two years making sure we had—from a client-customer perspective—a cloud-native, open-API, straight-through processing system.*

Ultimately, virtues like stability, reliability, and scalability are foundational qualities in the highly dynamic space of fintechs, new customer types, and innovative business models. The Bank of London has built a platform based on the robust, high-performance

SAP engine, which wasn't even designed for the financial clearing business, but is ready to take on clearing the trillions that make the global economy run.

6.4 Capital Markets: Giant Exchanges, Fintechs, and More

Grant shared his definition of capital markets and explained SAP's presence in the segment: "buy-side, sell-side, and market infrastructure of investment banking, investment management, and stock exchanges."

Even with SAP's large presence, Grant acknowledged its limitations. For example, SAP may not be a name brand easily recognized by capital markets organizations for our banking solutions, but the SAP portfolio includes software that is relevant to front, middle, and back offices:

> *There are two big areas where SAP has a footprint: First, we provide a range of corporate services and business support functions with our ERP solutions. But we also have a significant technology footprint. Global tier-one organizations on the buy-side and the sell-side have built significant securities trading and portfolio capabilities on SAP database, data management, and analytics solutions. Through the SAP Business Technology Platform, SAP provides tools for organizations to build their own applications on top. But increasingly, we're seeing that capital markets firms are looking at SAP's cloud solutions for operational capabilities such as finance and billing and revenue management.*

Grant provided a few other examples of trading exchanges that are big SAP customers:

- The London Stock Exchange has a sizable analytics presence with its acquisition of Refinitiv, the former financial and risk business of Thomson Reuters. With SAP, the London Stock Exchange has a data and analytics footprint with SAP HANA and with data intelligence. It has built a next-generation platform using our tools for generating sustainability indices and other index products—some of the biggest revenue-generating areas of its business.
- The Singapore Stock Exchange is using SAP SuccessFactors to change the way it recruits its workforce. In this region, many graduates don't want to work for a financial services firm. They want to work for a tech startup—something that is seen as sexy.

Many investment banks are using SAP solutions in creative "off-label" ways. JP Morgan Asset Management has embedded its liquidity platform directly into SAP S/4HANA, so corporate treasurers can invest that capital directly in the markets to improve efficiency. Rieker has called this approach "embedded finance":

> *JP Morgan wants to be at the desktop of the treasurer. They want to be at the desktop of the procurement officer. They want to be at the desktop when you book your travel. We're seeing more and more of these scenarios where the bank says, 'Can I embed my services into a travel expense system or travel booking system? Can I embed my services around payments and payment decisions?'*

Grant provided a wealth management example that illustrates customers shifting toward more autonomous functions:

> *The historical approach in wealth management was having several thousand advisors calling people. Today, people want to be left alone and rather use the digital tools to get information about what to do with their money.*

The Investors Group has used SAP HANA to aggregate data from many different systems, Grant said. It has combined and aggregated years of product, customer, and transactional data while maintaining the granularity of the data underneath. Then, the company has applied machine learning capabilities on this data to provide its clients an individualized roadmap for their financial lives:

> *You can log on as a customer and say, 'I want to retire when I'm 60,' and you go through an easy-to-follow process where you articulate your priorities. It will then basically suggest a plan to increase and structure your savings—by putting more into pension or investments or others.*

Clients end up with a personalized financial health and wealth journey with a digital user experience (UX) that is integrated with an intelligent, real-time decision engine harnessing more than 10 years of customer and product information. We find that this approach fits nicely in a society in which wealth management is no longer a service accessible only to the comparatively small population of the affluent elderly. Digital technologies and AI allow innovative wealth managers to address a broader population with lower wealth thresholds who are still building their careers or are not in line for a big inheritance.

Grant next turned to fintech startups and "unicorns,"[14] focusing first on fintechs' strengths at filling the operational gaps left by larger banks, such as payment speed:

> *Fintechs are very good at identifying weak spots in financial processes and homing in on them. They are also unwavering cloud-first entities. They don't want the data center baggage because many of the founders come out of larger financial institutions.*

We're seeing the biggest growth of fintechs in the payment space. This is especially true in retail, where the aim is to narrow the gap between consumer and retailer by providing that payment framework. It can still take up to three days for a transaction to be fulfilled from one end to the other. Fintechs build their business around fixing this inefficiency, something that larger banks have historically never really tried to tackle.

Grant's unicorn of choice is Mollie, a Europe-based transaction solution that is gradually spreading beyond its Dutch roots. Mollie maximizes the use of the existing frameworks that are already in place: Visa and Mastercard, PayPal, Apple Pay, and others. Mollie acts as an aggregator and provides a consistent transaction framework for retailers regardless of the underlying transport mechanism:

> *Then they'll start to build in other services, so that's where this kind of scalability element comes in for many of these fintechs. Most of them also want speed. They want to be able to set these systems up within three months.*

Mollie runs SAP S/4HANA in a public cloud. In the Netherlands, Mollie's home market, regulators have a rather open mindset in terms of public cloud solutions, which drastically changes how a financial system can operate. These ripple effects go beyond the Dutch borders into other areas of Europe.

Grant has seen fintechs redistribute power in the banking field at the digital cutting edge, while some larger banks still seem to overlook the virtues of the new digital age:

> *Fintechs will effectively expose and exploit the self-inflicted complexity that tier-one institutions have developed over the last decades. The big organizations are essentially balancing between keeping the lights on, leveraging the investment that they've made over the last 20+ years and then trying to adapt to the new environment where cloud computing offers innovation opportunities with flexibility and scalability. However, some of the more established players are still anti-cloud. They view cloud as witchcraft, and they don't want to go anywhere near it.*

Grant saw opportunities for tier-one institutions to use their power to contain fintech threats:

> *One advantage the bigger entities have is their balance sheets and the ability to acquire the fintechs and to bring them in if they start to become too much of a threat or if they have a complementary business capability that the large banks want to take advantage of.*

In this sense, fintechs are test subjects in an outsourced financial services innovation lab. The incumbents have the opportunity to pick thriving specimens and bring them into the fold—unless they miss the window of opportunity, and an 800-pound gorilla escapes the lab.

6.5 The Deutsche Börse Group Story

Deutsche Börse Group traces its history back to 1585, but its modern digital incarnation is only 30 years old. Within this short timeframe, Deutsche Börse has become one of the most influential international exchange organizations and market infrastructure providers. It claims that around 60% of the European energy trade flows through its platform, and that the same volume is true for commodities, foreign exchange, and more. Their biggest exchanges are 360T for foreign exchange and Xetra, the fully electronic trading platform where 90% of stock trading in Germany occurs.

Lars Bolanca, senior vice president and the head of corporate IT, explained how Deutsche Börse supports the full value chain of financial markets from pre-trading over trading and clearing to post-trading.

> *Pre-trading supports financial markets with high-quality, low-latency, and reliable data. Historical market data provides valuable insights for analyses, forecasts, and trading strategies. For trading, the group operates on-exchange markets for equities, exchange-traded products, bonds, and numerous other products. In addition, we also cover energy-related products, commodity products such as agricultural products and carbon dioxide certificates, and financial instruments such as foreign exchange, money market, or interest rate products. For clearing, the group operates two clearing houses, which act as central counterparties (as buyer to each seller, and as seller to each buyer) to minimize credit default risk—in simple terms, a kind of 'notary function in the middle.' Lastly, post-trading services cover settlement, custody, collateral, and liquidity management. Clearstream, Deutsche Börse's provider of post-trading services, is responsible for efficient global securities settlement and processes around 170 million settlement transactions each year.*

Grant emphasized a new challenge that exchanges face, compared to the more stable analog years of the past:

> *In the early days, exchanges operated in coffee shops and later became a venue for traders to meet and perform their transactions. In contrast, today's exchanges are a unique class of institutions. At heart they represent the epitome of a fintech. They use cutting-edge technologies within the financial*

services organizations, yet they have to operate within what is probably the most constrained regulatory landscape of financial institutions. They are the most systemically important element of financial markets.

But this role doesn't protect from competition: To stay relevant, exchanges must continuously innovate and support all the products that an exchange trades in a fair, efficient, safe, and secure way. To this day, they are lighthouse examples of how fintechs can and should operate.

Bolanca noted that Deutsche Börse's indices business supports nearly 800,000 structured products and is already looking ahead to reconciling tomorrow's trading patterns against a heavily regulated global marketplace:

“ *And in the future, you're looking at blockchain applications. How do you manage digital assets independent of borders, for example?*

Part of Deutsche Börse's challenge is to assess technology trends to stay ahead. It must analyze cloud computing, artificial intelligence (AI), distributed ledger technology and blockchain, big data, and cybersecurity to understand the potential and real impact on its business model. Quantum computing might have a strong impact while the Internet of Things (IoT), next-generation connectivity, algorithms on chips, and advanced processing units are expected to have a moderate impact. Deutsche Börse has developed centers of excellence (CoEs) to focus on high-impact trends, and it regularly assesses new trends to allocate its technology-related investments.

For example, for high-frequency trading areas, Deutsche Börse's computing resources are deployed largely on-premise, simply because hyperscalers cannot deliver on its data latency requirements yet. “Our trading environment is optimized at a nanosecond level,” said Bolanca, “I recently learned from AWS that the fastest round trip they can commit to is around 500 milliseconds, which, in the high-frequency trading world would be slow.” Considering that every millisecond has a million nanoseconds, this is the relative difference between one minute and two years. Meanwhile, he said, Deutsche Börse's mainframes are still the workhorses because they can support synchronous parallel processing. He considers that cloud computing can handle sequential processing and independent parallel processing at very high scale, “but cloud computing is not as good with classic T-accounting where all accounts must be balanced 100% in every nanosecond.”

When it comes to quantum computing, Bolanca said he's seen more traction in life sciences than in financial services. He has seen some pilots, but in the finance sector, with its more encapsulated services, not enough use cases exist yet for this very different computing paradigm.

Grant added another perspective about quantum computing:

> *The biggest opportunity for quantum is around risk analytics. However, a major concern is its potential for misuse by trading environments that don't want to play by the rules but rather seek the opportunity to game the markets in new ways by bending the rules and discovering new loopholes. Exchanges have a regulatory obligation to monitor the activities on their exchange for known illegal schemes. But they also have an obligation to monitor their market for undefined illegal activities. That's just a bottomless pit.*

Bolanca described the journey with SAP, Microsoft, and other cloud vendors starting in 2018, as shown in Figure 6.2.

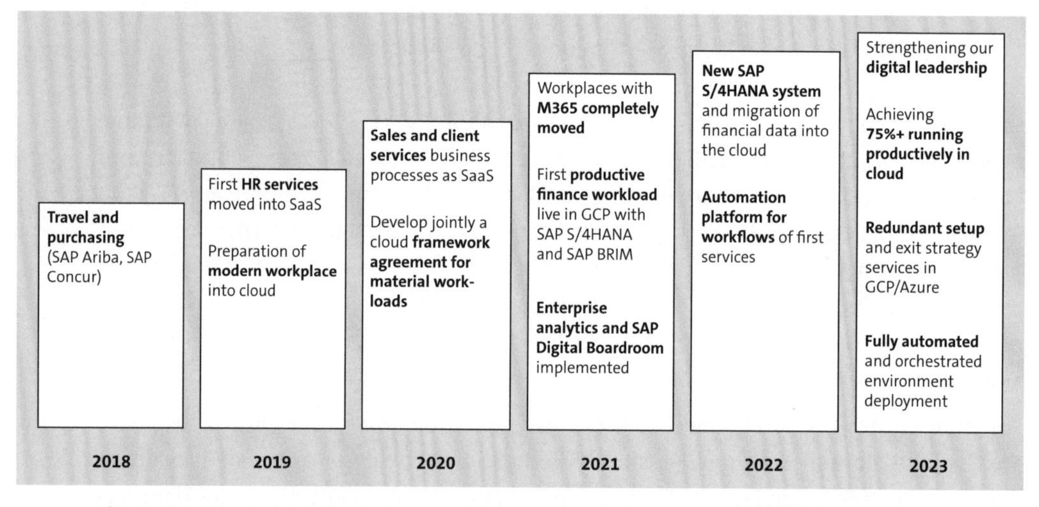

Figure 6.2 Deutsche Börse's Roadmap to the Cloud (Source: Deutsche Börse)

Back then, regulators were not as familiar with cloud computing and had concerns about security, data migration, data access, and data integrity. So, Deutsche Börse started with applications not directly related to the core business and not regulated in the finance sector, such as travel and purchasing. These initial steps allowed Deutsche Börse to experience applications in the cloud and on mobile devices, and regulators could monitor these early cloud experiences. Next was human resources and moving to Microsoft Office 365 in the cloud. In 2020 and 2021, Deutsche Börse started moving sales and client services to the cloud, which received more scrutiny from regulators. Deutsche Börse adopted frameworks from the German Federal Financial Supervisory Authority (BaFin). It already had workloads and parts of SAP S/4HANA running in the Google Cloud Platform, and since 2022, those have been growing. Deutsche Börse will be the first client running a full SAP S/4HANA environment—including side systems—in the Google Cloud Platform, with all the regulatory approvals across multiple jurisdictions.

Deutsche Börse is using SAP S/4HANA as its finance and treasury core, with integration to governance, risk, and compliance (GRC) functionality for group risk management in the areas of process controls, access controls, and master data governance. Per Bolanca:

> *We target having nearly everything in the cloud. We also want to monetize and better use our historic data. Last year we had over 900 terabytes of data only in the bigger warehouses, not even considering decentral data storage. And in our business we also have a stewardship role. In general, in an exchange, you prefer to keep data longer rather than shorter. But you also don't want to be sitting on thousands of terabytes that nobody will ever access.*

For Bolanca, the way forward is determining where it makes sense for Deutsche Börse to follow best practices and where it needs to adopt its own practices:

> *For our corporate services and processes, we follow what SAP is doing by moving as much as possible into software as a service. In parallel, we are standardizing corporate processes where Deutsche Börse is comparable to standard, such as purchasing and HR. Other corporate processes remain unique where we or our business model is special, such as financial risk management, billing, or treasury.*

Bolanca sees more possibilities ahead for process automation in customer interactions:

> *We have client services teams in the trading and post-trading area who handle thousands of customer interactions or orders every day with institutional customers. Here we see huge potential for further automation. Consumers today can open a bank account with a fintech in a few minutes, online, on their mobile device. In comparison, opening an institutional trading account can take up to six months. This is not just a technical problem. We are highly regulated, and are required to do comprehensive background checks on the institutions, their executives, or trading or transactional staff, etc.*

Having previously worked at SAP, Bolanca said he has been pleased to see SAP become a lot more familiar with the finance sector:

> *In our ambition to lead modern technology also in the corporate services world, we keep pushing SAP to upgrade to financial services industry needs when it comes to data protection, managing services in the cloud, audit trails, information security, constant evolution of conformity, real-time reporting on incidents or security breaches, and so on. This is something I would say has*

accelerated in the last two or three years. Prior to that, SAP appeared—especially due to acquisitions—to be mainly focused on the US market, with products like SAP Ariba and SAP Fieldglass. They have really made good progress in the last couple of years, but still have some way to go.

Grant agreed with Bolanca's candid assessment of SAP's growth in this space:

> *I think it is a fair statement. Historically, SAP was focused on core banking and core insurance areas. The exchanges, asset management, and investment banking are different animals. SAP has acquired a customer base through various acquisitions and we have a strategy in place.*

6.6 Insurance: Moving Way Beyond Premiums

Like every other industry, the insurance space is experiencing innovation from digital-first startups and tech giants who seek new markets for expansion. Tomic provided a perspective on the challenges the established insurance players have been facing:

> *We estimate that only 40% of insurance companies have a digital go-to-market strategy. They face competition from Google which is experimenting with auto insurance. Amazon has a partnership around small business insurance.*

But this challenge is nothing compared to Ping An out of China, as business analyst Simon Taylor explained on his Substack page:

> *The story is that Ping An is a tech giant masquerading as an insurance and financial services conglomerate. Annually, Ping An commits 1% of revenue—roughly 10% of profits—to R&D across its 'five ecosystems': financial services, healthcare, auto, real estate, and smart city. In another stroke of strategic clarity, they then cut those five ecosystems with three core technologies: AI, cloud, and blockchain.*[15]

Its Good Doctor service is merely one example of Ping An's huge impact, as Taylor described:

> *The company recruited a full-time in-house team of doctors with high qualifications to provide 24×7 online consultations and partnered with other healthcare players to build a 'closed-loop' system of healthcare services with fast delivery of medicine to patients' doorsteps and convenient access to partnered clinics and labs. Ping An Good Doctor has been addressing the needs of a vast population of Chinese people who are looking for high-quality basic care and*

health management services while creating a win-win between the company and other players such as hospitals, grassroots healthcare institutions, pharmacies, and independent clinics and labs.[16]

SAP is starting to see some dramatic moves among its customers. Reinsurance[17] is a traditional market for SAP, and reinsurance giants Munich Re, Swiss Re, and IRB Brasil all use SAP's reinsurance solution, part of SAP S/4HANA for insurance. This solution supports all forms of reinsurance—obligatory and facultative, proportional and non-proportional, assumed and ceded, life and non-life.[18]

In our discussion of the Integrated Mobility megatrend in Chapter 3, we discussed the impact of the CASE (connected, autonomous, shared, and electric mobility) paradigm on the automotive industry. As the car insurance industry feels the impact of trends in car fleet operators for mobility services, companies must debate the how the risks of autonomous vehicles should be distributed between owners, car manufacturers, and passengers.

Munich Re has of course been thinking about these changes. For this company, traditional auto insurance as we know it has already peaked and will be in steady decline for the next decade. We have all lived with insurance policies where premiums are largely influenced by our age and driving record and are locked in for months at a time. We may have even tried policies that use instrumentation to monitor our driving habits or charge us by miles driven, but the model has not evolved much so far, even though the industry is expecting change to accelerate.

Another big challenge in insurance is related to talent. Insurers have even more of a reputational challenge than banks, so they must be creative in winning and keeping talent. For example, Accenture is helping insurance giant AIG recruit AI experts.

On the subject of attracting and retaining talent in the insurance field, Tomic said SAP's talent management tools are attractive for the industry. SAP's sales compensation tool SAP Sales Cloud is quite popular and has proven especially successful with health insurers:

With more than 100,000 agents, a multinational insurance company from the US uses that solution to calculate sales commissions. We also help with agent licensing and training. In the US, you need to license agents by state and we help them with that.

Other aspects of running an insurance enterprise require an operational overhaul as well, and Tomic said he saw insurers finally coming around to the modern digital approach:

> *Five years ago, I could not get insurance companies interested in moving to the cloud. Now most are doing so because of the cost-efficiency aspect, but also because they realize the need to have standardized processes and standardized business models. Our RISE with SAP initiative enables several insurance companies to rationalize their portfolio of products and services.*

Tomic pointed out that the SAP Fioneer solution portfolio can run multiple insurance lines of businesses—life, health, property, and casualty—and their related end-to-end processes all on one platform, from policy management to claims. Particularly in Asia, many big insurance carriers run multiple lines of business, which fits nicely into the "super apps" we see across so many industries there. SAP provides a broad portfolio of capabilities spanning insurance and banking that addresses the needs of key business areas. SAP solutions for customer experience management support the chief marketing officer, while line of business heads need sales performance management. We have core insurance and core banking with SAP Fioneer today. And of course, we continue to support the CFO's office with our finance solutions and corporate services like procurement and real estate.

6.7 The ERGO Mobility Solutions Story

In our discussion of the Integrated Mobility megatrend in Chapter 3, we broadly considered the impact of CASE trends on the automotive industry. ERGO Mobility Solutions, part of the Munich Re family, was launched in 2017 to understand, influence, and capitalize on disruptive mobility trends. CEO Karsten Crede characterized the transformation in his LinkedIn essay:[19]

> *The automotive market is a massive growth area, which is very open to investment and innovation, but is also a very emotional market. It's different with insurance. The industry tends to be more cautious and more long-term oriented. How do these worlds fit together?*

ERGO's thinking has followed the CASE framework, seeing possibilities for multimodal, data-driven mobility that involves the following insurance-related features:

- Connected mobility supports pay-per-use insurance, connected assistance, digital and travel concierge services, and even digital claims.
- Autonomous mobility with advanced driver assistance systems and automated vehicles changes the risk structure and risk allocation.
- Car sharing, ride hailing, ride sharing, and mobility as a service use convenient, effective, and efficient insurance models that protect drivers, their passengers, and vehicle owners.

- Battery-powered vehicles carry specific risks and need additional products like battery warranties and insurance for widely distributed charging infrastructures.

At the 2022 SAP Fioneer conference in Amsterdam, Crede described how ERGO has progressed to the second generation of its business model.[20] He described his role and career focus as a "hybrid between the insurance and the automotive mobility industry" (he had previously spent time at Volkswagen Financial Services and at Allianz Insurance) and wants to position ERGO as an "insurance orchestrator in the mobility ecosystem."

ERGO's reinsurance business allows it to influence primary insurance offerings in many markets and to create solutions collaboratively with its automotive partners, who themselves have financing and insurance units.

In his talk, Crede contrasted the automotive industry, which moves fast and is hypercompetitive, with the insurance industry, which is heavily regulated and generally more conservative. This culture gap has spawned many contentious conversations, such as about who owns and who can access the data from connected vehicles. Crede suggested how insurance companies can do their part in bridging this cultural gap:

> *You need dedicated teams that have the automotive background and also know how to speak about financial services, and understand the strategic battlefields of the car industry, from sales to after-sales, and from innovation to scalability. We have a very clear understanding the car itself is getting smarter and that auto insurance also needs to become smarter. It needs to become more digital and needs to be integrated within the car software with over-the-air updates and the ability to buy car features on demand. We cannot continue with our 40 printed questions.... It's no longer the car you buy in the dealership (it's continually being updated) and the insurance provider needs to be prepared for this.*

He described how the "car of the future" is already influencing auto insurance and projected that talk of the electric drivetrain will last for the next decade. Meanwhile, discussions about autonomous driving levels are picking up speed. Daimler has a license for level 3 autonomous driving on the Autobahn, and there's already even a legal framework in place in Germany for level 4 driving. Consequently, Crede identified software and battery technology as future battlefields of the car industry. Forty percent of the value of an electric vehicle will come from its battery. Future profits will be tied to the software and the services that can be sold "over the air." The tech companies of Silicon Valley battle with the classic German car industry—and then there is China.

Industry giants are moving. Volkswagen has announced a 10,000-employee unit called CARIAD to help the company evolve from a traditional carmaker into a software company.

Crede explained that the insurance industry is aware of the impact of mobility trends on their business:

> *Different business models—auto leasing, rental, sharing, subscription—are evolving and blending. When it comes to leasing of electric cars, dealers may not be as influential, and this will change how you sell and service the car after the sale. When it comes to rentals, we all know the expression, 'You don't need to be gentle. It's a rental.' From an insurance perspective, that has been a disaster.*

But, per Crede, there's still opportunity for mobility providers and their insurers to distinguish between gentle and not-so-gentle drivers:

> *We need to better understand the profile of the person driving under each business model and come up with better individual pricing. We need to think about drive style analytics; it's much more than the classical telematics approach, which gives some benefit to careful drivers. Cars that are used only one hour a day [by one driver] are a different risk than shared cars that are and used for three to five hours a day [by different drivers].*

Crede's analysis also shows how different engineers and insurers look at the driver: He thinks about business opportunities for insurance, whereas engineers may agonize about automatically adjusting seat and mirrors for different drivers. But both elements will contribute to the new mobility experience. To focus on developing innovative automotive insurance products and risk management solutions, ERGO has announced opening a Mobility Technology Center with several partners.[21] Its initial focus areas include developing models for predicting the aging of battery cells under real operating conditions, which would allow providers of performance guarantees to offer insurance solutions for such batteries. The organization is also offering analyses of the safety and effectiveness of driver-assistance systems. The idea is to provide partners with a dynamic assessment of risks with a view toward guiding ERGO's customers to purchasing and using these assistance systems.

For Crede, collaborations like the Mobility Technology Center enable organizations to merge their key skills and knowledge areas with others:

> *From the insurance side we have some experts. But we would never be able to speak on an eye-to-eye level with the technical units of BMW or Volvo. The*

Mobility Technology Center gives us an 'extended workbench' with the technical expertise to—for example—find a sustainable insurance solution for a technical battery problem.

Figure 6.3 shows a topology of applications ERGO is deploying to help cover all these scenarios. SAP S/4HANA is the engine behind the box called the Motor Insurance Platform. The layer at the top contains the functions supporting digital interactions with customers, both in car and over better-established mobile and desktop channels. At the bottom, you'll find a true treasure trove of car and driver data that waits to be exploited by AI and used by smart processes. The middle layer contains modules that connect driver and vehicle data to support designing and selling additional services but also includes functionalities like claims management or collections and disbursements to link car and driver to policy and claims management, augmented by real-time data coming from the vehicle itself.

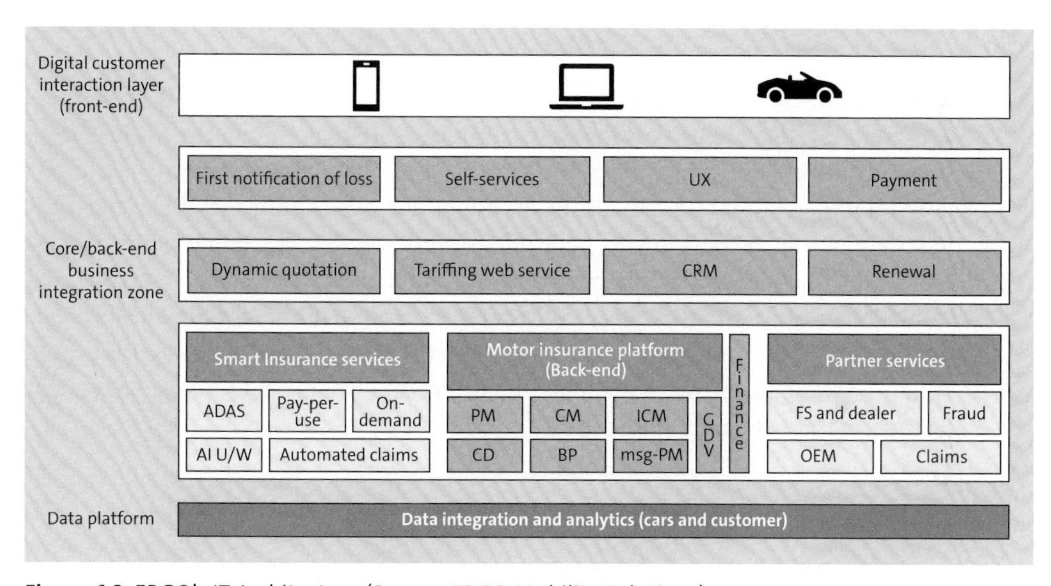

Figure 6.3 ERGO's IT Architecture (Source: ERGO Mobility Solutions)

While no industry sector has been spared the shocks of digital transformation, COVID-19, and the Ukraine crisis, you could argue that financial services sector has been exposed to disruptive change for much longer than others. While SAP continues to be viewed more as an industrial software vendor, our wide customer base includes specialty banks, capital markets, and innovative insurance companies who are embedding themselves in a wide range of ecosystems in other industries.

Grant noted that "the financial services sector has the potential to be the catalyst for change and to support a lot of the other things that are reshaping every other industry."

6.8 The Future of Capital and Risk with SAP

Many of us will perceive financial services somewhat removed from our daily lives in the "real world." If you casually walk into branches of banks you're not affiliated with to find out what they offer, you're a member of a very small minority. This level of detachment may seem strange because every business transaction around a product and service is linked with a financial transaction and, in many cases, with a reallocation of risk.

Embedding financial services in the daily lives of consumers and enterprises is a proven approach to creating "sticky" services that customers use because they are convenient and at their fingertips. However, providing financial services is no longer the monopoly of the financial sector, although financial services continue to be subject to regulation. Financial service providers respond to new competitors by developing new, fully digital services that fulfill the new needs of their corporate customer base. This trend is amplified by a personalization of financial services to individuals over their lifetimes: If you can buy personalized granola and get personalized medical treatment, you can also expect personalized financial services.

In the business domain, financial service providers aspire to become strategic partners who co-innovate new business models and advise on risk allocation. The world of capital and risk is continuously changing and subject to sometimes violent market forces, political intervention, and the shifting of geopolitical and cultural tectonic plates. Capital is seeking sustainable investment opportunities under a broad definition of sustainability that covers ESG performance, product portfolios, global markets, and technology trends. In a volatile and dynamic market environment, financial service providers need the best available talent, and thus, routine work is being handed over to systems powered by artificial intelligence (AI). Recruiting and retaining the right talent has become a top business priority.

6.8.1 Embedded Financial Services

As the financial services industry expands and digitally transforms with fintechs and insurtechs taking the lead, the need for seamless integration between financial service providers and their customers are now table stakes.

The following solutions support embedded financial services:

- SAP Business Network embeds financial services, such as payment or financing, directly into SAP solutions as well as other organizations' products and services for a seamless workflow.
- SAP Business Technology Platform enables efficient integration of financial services in other business transactions, for example, in the area of commerce and trade finance. Process extensions and new analytics can be deployed while keeping the semantics of your business data intact.

- Business process analytics and modeling with SAP Signavio identifies and models process optimization, for example, between enterprises and their financial service providers.
- Qualtrics XM generates a continuous stream of user and customer feedback to identify opportunities to optimize processes and to adjust financial services to better meet the core needs of customers.
- SAP Multi-Bank Connectivity is an integrated network allowing financial institutions to directly connect to SAP S/4HANA Finance users, resulting in increased value and services through a seamless user experience.

6.8.2 Industry Convergence

Organizations across banking, capital markets, and insurance need to embrace collaboration between incumbents, new non-traditional business partners, and fintechs/ insurtechs to achieve geographical mobility and to create new business models.

The following solutions facilitate cross-industry collaboration and new business models:

- SAP S/4HANA Finance's treasury management and governance, risk, and controls functionality gives organizations agility and the ability to meet the requirements of a new jurisdiction when expanding into new regions or when evaluating and starting new businesses.
- SAP Billing and Revenue Innovation Management enables firms to implement subscription-based business models, analyze product performance, and quickly implement new business models with native integration with SAP S/4HANA for cash management and invoice transparency.
- SAP Configure Price Quote allows organizations to build customer-centric sales activities with the agility to adapt to changing market needs.
- SAP Business Technology Platform enables integration across on-premise, private, and public cloud environments with rich application extension capabilities and on-demand, self-service analytics.
- SAP Signavio enables fast process alignment and the modeling of innovative processes to support new business activities across companies and industries.
- SAP Business Network offers organizations the ability to source unique business insights from SAP and partner business networks, thus driving new intelligence and value-adding analytics.

6.8.3 Attracting Talent

As financial services organizations embrace digital transformation and strengthen their business and technology capabilities, they are facing a constant battle to attract and maintain the right talent while ensuring that their training needs and regulatory obligations are met. This battle makes having the right in-house talent and external talent more important than ever. A carefully designed "employee experience" from finding and recruiting talent through onboarding, training, and career development is key to shape and retain a high-performance workforce. Successful financial services institutions take a broad look at their own employees and contingent labor to run a workforce that is qualified and motivated to make a difference for the business.

- SAP SuccessFactors supports the transformation of organizational culture from hire-to-retire to one that attracts new talent. The solution also delivers analytics to assess and appropriately manage existing talent and to provide relevant training capabilities that meet employee and regulatory requirements.
- Qualtrics XM deeply analyzes detailed employee and agent/producer feedback to use the insights to optimize the employee experience.
- SAP Sales Performance Management and SAP Agent Performance Management help maximize sales performance for banks and insurers. These solutions cover licensing; training and onboarding/offboarding of agents, brokers, and producers with automated processes; intuitive user interfaces (UIs); and impactful incentive and compensation management.
- SAP Fieldglass is an open vendor management system (VMS) network that helps organizations find, engage, manage, and pay external workers.
- SAP Business Technology Platform supports the integration of SAP to non-SAP landscapes and external systems—for example, to the providers of temporary labor—with unifying analytics to support on-demand and self-service business analysis.

6.8.4 Sustainable Financial Services

The focus on how capital is invested has never been stronger: Private and institutional investors seek out investments in sustainable enterprises, so favorable sustainability ratings have become critical for access to capital. Organizations now need granular analytics and detailed records about their use of capital across financial products, ranging from consumer loans to portfolio investments and insurance policies. ESG data must be central to the financial product development process as well as by supporting the analysis and performance of existing products. Asset managers face the challenge of racing to net-zero emissions in their portfolios while banks and insurance companies try to bolster their reputations by ensuring their products and services are in line with society's ESG expectations.

- SAP Business Technology Platform allows organizations to acquire and process the multitude of complex datasets needed to support ESG analytics using document processing and AI for data quality routines.

- SAP Profitability and Performance Management executes the complex granular analytics with embedded ESG taxonomies to support investment processes combined with traditional portfolio management analytics.
- SAP Sustainability Control Tower provides a corporate reporting framework and analytical tools to understand the organization's position against key ESG metrics.

6.8.5 Finance as a Strategic Partner

The future success of organizations will depend on their agility to drive business outcomes. It is no longer sufficient for finance to still be receiving data with a 2- to 3-day delay; data needs to move continuously between front and back office with minimal or no intervention or reconciliation holding it up. Finance personnel need on-demand, self-service access to information for their business activities. They need support from a system that enables a modern target operating model and uses machine learning and AI capabilities for advanced automation and decision support.

The following solutions help position finance departments as a strategic partner to the business:

- SAP S/4HANA Finance allows organizations to act on the latest data for improved financial planning, capital analysis, and group and regulatory reporting, through simplified accounting and closing practices and automated workflows.
- SAP Business Technology Platform enables data from front- and middle-office systems to be consistently and continuously integrated with SAP systems of record for internal and regulatory reporting by facilitating true straight-through-processing. Advanced analytics can natively be embedded in the data as it moves through processes to provide additional insights.
- SAP Profitability and Performance Management defines and monitors profitability by line of business, branch, and banking or insurance product to the lowest level with the full power of cost allocation. This solution provides input into future pricing.
- SAP Financial Asset Management provides a 360-degree view on investment accounting and manages capital management across the board for group corporations in different countries with several parallel accounting systems. This solution enables better control and tighter management of investment portfolios.
- SAP S/4HANA's group reporting functionality unifies financial closing processes and eliminates the need for multiple systems by drilling down from consolidated reports to see a document's detail in the general ledger.
- SAP Signavio allows firms to eliminate manual reconciliation processes, resulting in seamless integration of data from downstream systems to SAP S/4HANA, which can then be continually enhanced to eliminate error sources and gain further efficiency over time.

- SAP Collections and Disbursements for Insurance allows organizations to perform all collection- and disbursement-specific tasks across lines of business (such as open item accounting, payment processing, incoming payment processing, correspondence, and dunning) and also allows you to map invoicing that involves brokers or co-insurance businesses.

6.8.6 Personalized Financial Services

Customer expectations have drastically changed through interactions with global consumer technology companies. Financial services organizations are not immune to this shift, and consumers expect more tailored and personalized experiences through a variety of interaction channels, with 24/7 availability.

The following solutions that support hyper-personalization might be part of your SAP-run IT landscape:

- SAP Customer Experience combines customer data with marketing, sales, commerce financial service accelerator for banking and insurance and Qualtrics XM to provide a unique capability to understand and target individual customers in the way they want to be reached.
- Personalization of products and services at scale requires deep business process insights. SAP Signavio enables individualized service without excess effort by replacing or digitally augmenting manual processes and continuously refining digital processes to improve outcomes and efficiency.
- SAP Business Technology Platform provides integration, database, and analytics capabilities to quickly connect SAP and non-SAP sources, allowing business users to enable self-service and up-to-the-second insights.
- SAP Fioneer products and services focus on accelerated innovation in banking, insurance, and software solutions specific to the financial services industry. They enable customers to innovate and transform their businesses by meeting the need for digital innovation and cost efficiency in financial services.

Chapter 7
Sustainable Energy

The Struggle with Finite Resources and Climate Change

7

Your body needs energy just to stay alive, and evolution has made our bodies surprisingly energy efficient: If you were powered by electricity, you would need a daily charge of about 2.4 kilowatt hours (kWh), the equivalent of some 2,000 kilocalories. Check your utility bill: Even if you pay a lot for your electricity, your body could probably run on less than a dollar a day. Granted, enjoying a nice dinner sounds more pleasant than being plugged into a charger overnight.

Compared to the little energy we need to live and breathe, we use a lot of energy every day to move around, to stay warm, to communicate, to make things. While energy is so essential for us, surprisingly few of us understand how the global energy system works and what levers we can pull to make this system more sustainable.

Over many millennia, humankind was living sustainably in the sense that we could have gone on forever without getting into trouble: We used our own bodies and beasts of burden to move around, and we got our energy from eating plants and animals and from burning wood—resources that would biologically be replenished. At this level of consumption, all greenhouse gas emissions were short cycle: The carbon dioxide emitted from burning wood was recycled when the next tree grew. So, humankind was carbon neutral. For hundreds of millions of years, nature was even carbon *negative*: Organic matter was trapped under ground and slowly converted into coal, oil, and natural gas, so the carbon dioxide didn't return to the atmosphere.

Early machines used renewable energy: antique watermills in Greece, windmills in Persia and northwestern Europe, and of course animal or human labor. So, until recent times (recent on geological timescales), people either weren't around or didn't mess with the atmosphere. Figure 7.1 shows that, over the past 800,000 years and many ice ages, the concentration of carbon dioxide in the atmosphere has oscillated between 200 and 300 parts per million (ppm). Just a blink of the eye ago (in 1950), the carbon dioxide concentration exceeded the former maximum and has continued to shoot up ever since.[1]

We are struggling with two distinct, but connected, problems: The high concentration of carbon dioxide in the atmosphere drives climate change through rising temperatures, and we are depleting the hydrocarbon deposits nature has accumulated over hundreds of millions of years in just a few hundred years. The impact of climate change on nature and humankind is becoming increasingly visible and tangible through extreme weather patterns—in floods and droughts, in melting glaciers, in the rising sea levels of our warming oceans.

Using oil and gas will deplete the planet's reservoirs over time, even if we continue to discover new oil and gas deposits. We already need more sophisticated extraction technologies: In early rushes for oil, all you had to do was poke a hole in the ground and install a tap. Today, we need hydraulic fracturing and horizontal drilling and cathedral-sized offshore rigs. Using oil and natural gas as fuels can be considered a barbaric use of hydrocarbons that have enabled the huge field of organic chemistry to make the paints, plastics, and performance materials that shape our world. As the famous 19th century chemist Friedrich Wöhler said:

> *Organic chemistry just now is enough to drive one mad. It gives one the impression of a primeval, tropical forest full of the most remarkable things, a monstrous and boundless thicket, with no way of escape, into which one may well dread to enter.*[2]

Carbon (C) is a truly polyamorous atom that loves to engage with most of its peers in the periodic table to form about 10 million known molecules—compare this number to about 100 thousand known inorganic molecules. We believe that leaving hydrocarbons in the ground for generations to come is a good idea so that future generations can still find resources enough for more creative and sophisticated applications than just using burning them as fuels.

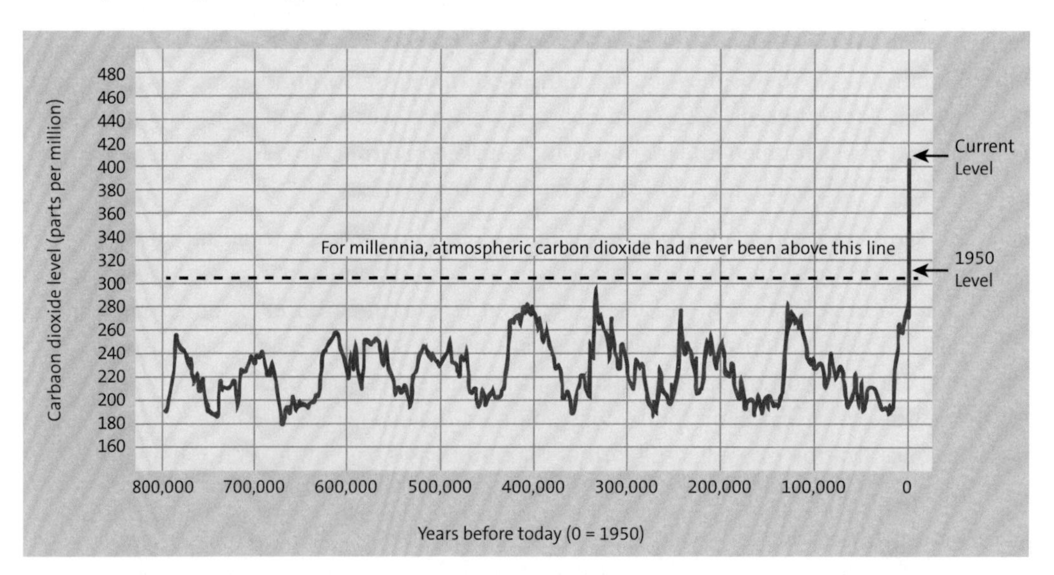

Figure 7.1 Carbon Dioxide Concentration in the Atmosphere over the Last 800,000 Years (Source: NASA)

Of course, there are less philosophical and more short-term and hands-on incentives to reducing our consumption of hydrocarbons for energy use. For example, the dependence of Germany—the fourth largest global economy—on cheap Russian gas and oil has shown the utility of hydrocarbons as geopolitical weapons. Russia is not the only nation with enough oil and gas to hold hostage countries with big demand and limited local resources.

7.1 The Heterogeneity of the Energy Sector

As with the other industries we discuss throughout this book, the "energy" industry has been undergoing massive changes. Benjamin Beberness, vice president and global head of the oil and gas industry business unit at SAP and a former chief information officer (CIO) in the utility industry, explains the different understandings of the word "energy":

> *In the utility business, when you talk energy, you're thinking of electricity and of the natural gas that we pipe to households. In the oil and gas industry, by contrast, when you talk energy, you're talking mostly about energy from fossil hydrocarbons, about generating renewable energy with wind and solar, and about hydrogen to store and move renewable electricity over space and time.*

Let's go back to Physics 101 where energy is defined as "the ability to do work." Fossil fuels can do a lot of work: 1 liter (about a quart for our imperial unit friends) of diesel contains about 10 kWh of energy. If you pedal on a bicycle ergometer, you'll probably sustainably output some 100 watts. So, if you attach a little generator and a battery to your bike and pedal for 10 hours, you'll have put a single kWh into your battery. Repeat for 10 days (2 weeks if you take the weekends off), and you would have generated the energy content of a single liter/quart of diesel.

Stephan Klein, senior vice president of SAP Customer Success, commented on how the discussion and concerns about sustainable and renewable energy span across industry boundaries:

> *Earlier this year, I was quite surprised when the organizers of the World Utilities Congress in Abu Dhabi asked me to speak about whether liquefied natural gas (LNG) is a transitional fuel or a long-term component of the energy mix. Previously, LNG would be a recurring topic in oil and gas conferences, and a utilities event would have me speak about renewable energy production or the digital prosumer. This time, they wanted to discuss the role of hydrogen and chasing zero emissions. Utilities or oil and gas conference—different attendees, but the same big topic: the sustainable energy transition.*

These concerns go even further. In July 2022, the European Parliament backed European Union (EU) rules labeling investments in gas and nuclear power plants as climate friendly.[3] The new rules will add gas and nuclear power plants to the EU taxonomy rulebook from 2023, enabling investors to label and market investments in them as green. This concession would have been unthinkable before the Ukraine crisis but shows either pragmatism or ignorance, depending on whom you ask. Of course, all the EU labels and taxonomies won't change the simple fact that burning natural gas (CH_4) will follow this equation: $CH_4 + O_2 \rightarrow H_2O + CO_2 + \text{energy}$. In plain English: If you burn methane, you'll get water, carbon dioxide, and heat. And yes, you can go ahead and capture the carbon dioxide through carbon capture and storage (CCS), but then, you can also do that when you burn coal or any other organic material.

Discussions about sustainable energy are not limited to Europe. Darren Woods, CEO of ExxonMobil, spoke about climate change and ExxonMobil's business in a 2022 interview with CNBC:

> *Two years ago, [the conversation was all] wind and solar, period. Today, it's wind and solar and carbon capture and hydrogen and biofuels and ammonia. So already, in the course of a very short period of time, there's a much broader recognition that there are a number of solutions required to make this thing work.*[4]

Accenture has long been a strong SAP partner in the oil and gas and utility industries. Muqsit Ashraf, CEO of Accenture Strategy, has been helping clients plan for energy transitions. He added a historical perspective on such transitions:

> *The world has been in an energy transition for centuries. We moved from biomass to coal to oil to now a multi-fuel energy mix. The two distinctive attributes of the current transition are significant demand-side changes required alongside supply-side shifts, and the convergence of multiple sectors needed to enable the transformation of the energy system.*

We are moving from single dominant energy sources to an energy mix of multiple sources. In its publication *The Changing Joule Dynamic*, Accenture noted the transition from the predominant use of wood to coal in the early twentieth century, followed by oil taking over in the 1960s.[5] Since then, coal and oil have been joined by gas, nuclear power, hydropower, and wind and solar power in the market (with biomass such as wood sources on the decline). The paper doesn't identify a dominant energy source: "No single winning fuel in the medium-term future. Uncertain paths for all energy sources." Definitely an interesting analysis. Keep in mind that Accenture's publication focuses on the changing energy mix and shows a drop of coal and oil in the energy mix, but in absolute numbers, the total energy demand has still skyrocketed from 7,300 terawatt hours (TWh) in 1840 to 159,000 TWh in 2021—that's a factor of 22.[6]

7.2 Global Energy by the Numbers

Data is not in short supply in the global energy market, but, ironically, plenty of misinformation and wishful thinking abound in the midst of this ocean of data.

Let's start with an analysis by Det Norske Veritas (DNV), a Norwegian organization founded in 1864 with a long history of cataloging maritime, energy, and other data. DNV's 2022 Energy Transition Outlook illustrates changes in energy consumption and in the energy mix over time.[7]

As shown in Figure 7.2, DNV anticipates that energy consumption will peak at 617 exajoules in 2030 and then slowly decline to 590 exajoules in 2050. In that window, it projects, the share of renewable energy such as wind and solar will increase; hydropower, bioenergy, geothermal, nuclear fuels, and natural gas will stay flat; and the consumption of oil and coal will decline. By 2050, DNV anticipates that renewable energy (wind, solar, biomass, hydropower) and fossil energy (oil, gas, coal) will be of similar orders of magnitude. We won't be the judge of where nuclear energy falls in this mix.

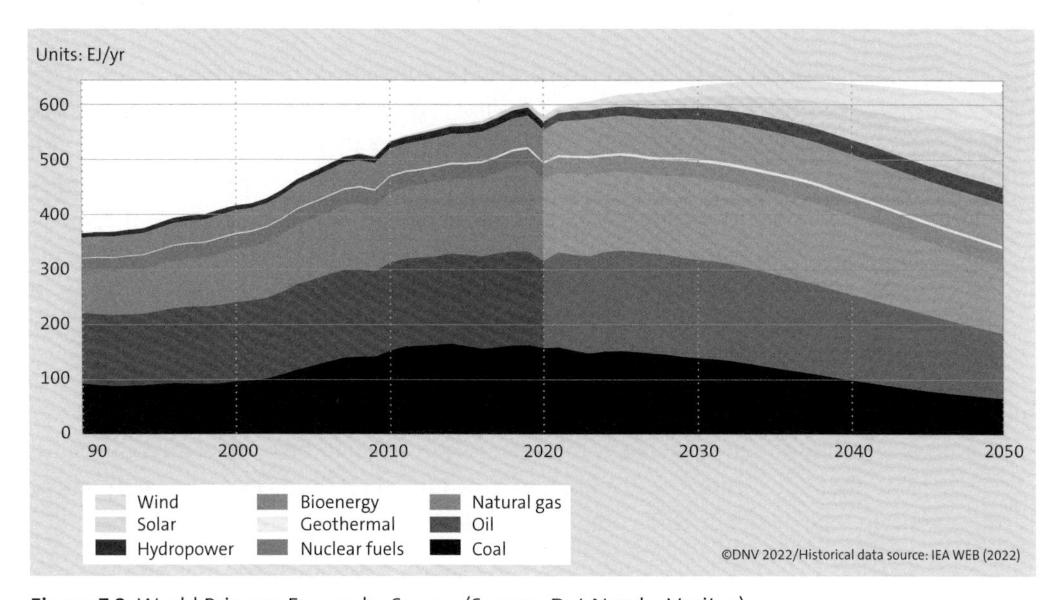

Figure 7.2 World Primary Energy by Source (Source: Det Norske Veritas)

Germany has been doing a lot of inventing and investing in the wind and solar energy space, triggered by the oil shock of the early 1970s. In the late 1970s, the wind turbine GROWIAN was developed, with test runs starting in 1983.[8] With a hub height of 100 meters and a rotor 100 meters in diameter, GROWIAN held the world record as the biggest wind turbine for many years. But it never worked properly, and it provided engineers with more lessons learned than energy to the grid. Today, GROWIAN would be a dwarf among the wind turbines now found in onshore and offshore wind farms that come with rotor diameters of 240 meters.

Despite the research and investment in renewable wind and solar energy, those renewables barely register as a global energy source today. But per Remi Erikson, CEO of DNV, wind and photovoltaic solar energy will expand 15- and 20-fold, respectively, in DNV's forecast period until 2050. "Twinned with the plunging costs and advancing technology of battery storage, variable renewables are already enabling a phase out of thermal power generation and the business case will become overwhelming by 2030," he wrote.

Accenture's Ashraf similarly noted a fundamental change that has occurred in recent years and has made solar energy more mainstream:

> *We are able to talk of solar today as a very competitive technology. That's a sea change that has only happened in the last decade. Solar panels have been around for a very long time for niche applications. But it wasn't until China made a conscious decision to heavily invest in solar in the last decade that that solar really started to actually beat the adoption projections. This suggests that the S-curve for energy technologies is likely to be steep, but to reach that point of fast growth, minimum scale in investments and demand is needed.*

We find that it's not trivial to compare the cost of a kWh from renewable sources versus fossil fuel sources, but the approach of *levelized cost of electricity* (LCOE) by source is a solid approach. Simply speaking, LCOE combines the capital cost for a power plant with the cost for its fuels over the lifetime of the plant. The findings of recent global studies vary somewhat, but they agree that (onshore) wind and solar have achieved the lowest LCOE compared to natural gas, geothermal, coal, and nuclear energy. This is not a big surprise: The cost of solar panels drops with the scale of production, and the output of bigger wind turbines is tied to the square of the rotor diameter (well, for energy experts: yes, it's a bit more complicated than that).

Peter Koop, global lead for energy transition for the energy and natural resources sector at SAP, has watched the renewables sector mature since he worked on his thesis at the Fraunhofer Institute for Solar Energy in the mid-1990s. Koop explained his point of view that solar power disrupts other energy carriers:

> *One was a large-scale power solar plant in Chile. The winning bid came in at $0.011 per kilowatt-hour. That's full production cost. Another was in South Africa. The tender was for a gigawatt, guaranteed to be available from 5 am to 9 pm. The bidders could bid nuclear, coal-fired, gas, renewable fuel, whatever—no constraints. The winner was a solar and battery combination, so a large solar farm and batteries to store the electricity for use in the morning and in the evening. That was the cheapest, most reliable combination. That*

says everything. For new projects, renewable energy plus batteries should be the solution to consider.

The solar panel plus battery combination holds a lot of appeal, but we must keep an eye on the new dependencies on large-scale sources of solar panels, batteries, and raw materials to make them. China is by far the largest producer of solar panels, and geopolitical considerations suggest that onshoring and near-shoring solar panel production might be a good idea. Making solar panels also uses a lot of energy before they can start collecting free energy from the sun. Studies seem to agree that it takes "a few years" until a solar panel has generated the energy that was used to make it.[9] With a solar panel life expectancy of 20+ years, that looks like a fair deal, in particular, if the energy used for solar panel production comes from renewable energy.

DNV's analysis considered a 2050 scenario in which about half of the world's primary energy demand is satisfied by renewable energy. Current numbers from Our World in Data say that solar and wind energy account for 4.6% of primary energy, as shown in Figure 7.3.[10]

DNV's projection believes that many gains from solar and wind will come at the expense of fossil fuels, but coal, oil, and gas are still expected to contribute roughly half of global energy needs in 2050. These numbers may surprise some readers—in particular in Germany—who read that 48.5% of electrical energy came from renewables in the first half year of 2022.[11] This is an impressive number, particularly if you consider that Germany isn't the sunniest or windiest place on the planet. But it's easy to overlook the fact that this number only shows the renewables share of *electrical* energy, not of *primary* energy. Focusing on primary energy, the contribution of renewables drops to 16% in 2021.[12] There is still a long way to go for renewable energy even in a country that takes pride in the *Energiewende*.

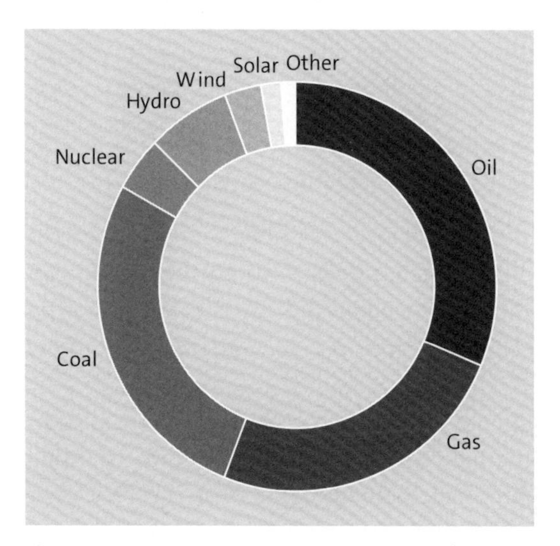

Figure 7.3 Primary Energy Sources in 2021 (Source: Our World in Data)

Daniela Sellmann, global vice president and head of the utilities industry business unit at SAP, cites another nation's use of primary energy sources:

> *Iceland already started its green journey in the 1970s due to skyrocketing oil prices and its respective reliance on this resource at that time. Thus, the country has been amongst the first to invest into geothermal energy, and is now acting as worldwide sparring partner for other utilities on their way to a greener future. Today Iceland's main primary renewable energy comes from geothermal (66%), with hydro in second place. Iceland is a role model with 85% of the total primary energy supply being derived from domestically produced renewable energy sources. This is the highest share of renewables in any national total energy budget. The remainder 15% are mainly due to transportation and shipment.*

A good example for substituting burning fossil fuels with using (potentially) renewable electricity comes from the transition to e-mobility, which is expected to add significant strain on the power grid, so the adoption of electric vehicles (and any other large-scale replacement of fossil fuels with electricity) needs a commensurate upgrade of the power grid. The need to balance energy demand and supply is nicely illustrated by an exchange in the US Congress between Representative Thomas Massie and US Secretary of Transportation Pete Buttigieg.[13] Massie stressed that, while he is personally bullish on technologies using renewable energy, "the numbers and the rate of adoption has been developed using political science, not engineering," and that current infrastructure is not prepared for a fast transition, which would result in rolling blackouts and brownouts. Buttigieg responded that power generation needs to be in lockstep with increasing demand, and that he doesn't want to give up because of the need to transform the energy system.

With his behind-the-scenes knowledge of utilities, Beberness has a more optimistic perspective than Massie that electric grids can handle electric vehicles, based on utilities companies' own research, but cited a lack of information sharing as one obstacle to public acceptance:

> *Utilities have been studying electrification of vehicles and the impact on the grid for at least 15 years now. How will their distribution network handle that increase? Which cities, towns, and neighborhoods will adopt EVs and at what rate? And what does that mean to their overall demand and supply patterns?*

While utilities understand these scenarios and the required response, Beberness saw a lack of aggregating and sharing data on a regional level. He also acknowledged that shifting loads from burning fossil fuels in vehicles and homes to electric vehicles and electric heating won't be simple:

> *Another concern is the amount of copper required to meet the demand of the energy transition. Some studies are showing that we would need to see copper production double by mid-2030.*

Beyond the new load placed on power grids, yet another concern is the range of relatively scarce minerals required for an energy system moving to renewables. For example, do we have enough lithium, nickel, cobalt, and other minerals for the batteries and motors needed for electric vehicles and for power storage? The CEO of automaker Stellantis, Carlos Tavares, has warned:

> *The speed at which we are trying to move all together for the right reason, which is fixing the global warming issue, is so high that the supply chain and the production capacities have no time to adjust.*[14]

Simon Michaux, research professor of geometallurgy from Geological Survey of Finland (GTK), gave an even more cautious perspective in a report:

> *The mass of lithium-ion batteries required to power the 1.39 billion EVs proposed in Scenario F would be 282.6 million tons. Preliminary calculations show that global reserves, let alone global production, may not be enough to resource the quantity of batteries required.*[15]

As we discussed in Chapter 3 on the Integrated Mobility megatrend, we may not even want to replace all the cars on the street one-to-one with battery-powered vehicles. So, maybe the huge lithium demand can be reduced by moving to new, integrated mobility.

Wind and solar are the most scalable renewable energy sources with their major drawback being that they are intermittent. You need a way to store sunshine during the night and wind power during lulls. Lithium-ion batteries are certainly not the answer, according to Michaux:

> *In 2018, pumped storage attached to a hydroelectric power generation system accounted for 98% of existing power storge capacity. The mass of batteries required is enormous. The estimated mass is 2.78 billion tons of Li-Ion batteries (where stationary power buffer storage makes up most of this). This estimate is so enormous, it becomes now appropriate to ask is it even possible in context of mineral reserves available, as this far exceeds global reserves and is not practical.*

In discussions about storage for intermittent energy sources like wind and solar, hydrogen is increasingly popular as a storage technology since it doesn't depend on equally scarce resources as battery storage. We'll come back to hydrogen later in this chapter.

7.2.1 Energy Transformation and Geography

For now, let's go back to DNV's research showing the projected energy mix in 2050 and to Accenture's analysis: Both suggest the need for a system-wide transformation of the energy system. Simply replacing gas-guzzling cars one-to-one with electrical vehicles will overload the grid if the grid doesn't adapt. Grid-level battery storage for 100% of intermittent wind and solar energy is certainly neither practical nor affordable. The energy system of the future will be diverse. As DNV predicts, hydropower, nuclear, and geothermal energy will continue to play small but varied roles on a global scale even in 2050. But preference for and access to these sources is often regional. For example, France and several other EU countries will continue to use nuclear energy as an important source of electricity. Growing yet cautious optimism surrounds small modular reactors (SMRs) that can be slotted into brownfield sites in place of decommissioned coal-fired plants, but we sit on the fence here, at least until the technology has proven itself. For Norway and Brazil, both blessed with plenty of deep valleys and rivers, hydroelectric power will continue to be the dominant energy source. For Iceland, Kenya, and the Philippines, geothermal energy is critical. All these energy sources go through their individual waves of innovation—like supercritical geothermal power[16] and gravity-powered energy storage[17]—but be careful, noise and nonsense abound in the storage space: Pumped-storage hydroelectricity, discussed in detail in Section 7.4.3, is a great and proven technology but with a limited set of viable locations. Other proposed and prototyped energy stores require a little tire kicking: Use $E = m \times g \times h$ (energy equals mass times the gravity constant times height) to calculate how much potential energy you can store by lifting stuff up and later recovering the energy by lowering it again. More often than not, this back-of-the-envelope math is sobering—and can keep you from investing in startups that are predictably duds.

Woods of ExxonMobil commented on the geographical distribution of energy sources, and emphasized his perspective that the main objective is lowering greenhouse gas emissions (as opposed to solving the problem of finite energy resources):

> *What we'll find as we go around the world, that the natural endowments, the advantages that you see, the technology progress that you see is going to manifest itself differently. And we shouldn't care, as long as it's achieving the objective, which is to lower the emissions.*

This global perspective focuses on the geographic viability of different forms of renewable energy. A look at a globe or at Google Earth shows that the best areas for solar energy are deserts, and the best wind resources can be found offshore and on mountains. Unfortunately, these geographic regions generally do not host population centers and industry hubs, so the question remains of how to transport energy from remote regions to the locations where the energy is needed. Now, overlay the globe with a geopolitical map, and you'll find that many deserts and mountains are situated in countries that are well aware of how big a bargaining chip energy can be in geopolitical power struggles.

7.2.2 Energy Transformation and Industries

An energy system in transformation impacts different industries differently. Some industries can make the same product with different energy types: The steel industry now has the opportunity to use preferably green hydrogen instead of fuel coke. Furnaces can run on renewable electricity instead of gas. As discussed earlier, electricity can be generated from a broad range of renewable energy sources. The chemicals industry can replace fossil fuels with hydrogen and synthetic hydrocarbons. The cement industry accounts for a whopping 8% of global carbon dioxide emissions, but they can use hydrogen as a fuel and CCS to deal with the carbon dioxide emissions from calcination reactions. Agriculture runs on ammonia-based fertilizers. Today, ammonia is made from natural gas, but it could also be made from green hydrogen. And of course, utilities can make electricity from renewable sources, and the oil and gas industry can make synthetic hydrocarbons from carbon dioxide, hydrogen, and renewable energy (pretty much reversing the process of burning hydrocarbons).

So, if all the industries can so easily be decarbonized—why don't we just do it? Well, the capital costs to replace current production assets are astronomical, the required technologies are still immature, and even if a genie could replace the capital assets worth trillions overnight with still-to-be-developed new technology, we would still lack enough supply of renewable energy and feedstock to power the new production assets.

Figure 7.4 shows the scale of the problem: To make oil and gas production, steel, cement, aluminum, and ammonia carbon neutral, we need to scale carbon capture, transport, and storage capacities by a factor of 63. Clean hydrogen must grow by a factor of 8, and to make the industry carbon neutral would require almost half of the clean power we have available today.[18] That's why we talk about transforming, not swapping, the system.

Other industries need to make different products, not just the same products differently. The automotive industry is in the middle of the transformation from internal combustion engines and fuel tanks to electrical motors, batteries, or even hydrogen—the former a mature technology, the latter two still research and development in progress. Industrial machinery and components and engineering and construction industries have to provide the requisite genie with all the shiny new equipment to run on or make renewable energy. The mining industry faces a steep decline in the demand for coal but would see big demand for a wide range of minerals required to make solar panels, electric motors, more electronics, new catalytic converters, filters, and whatnot.

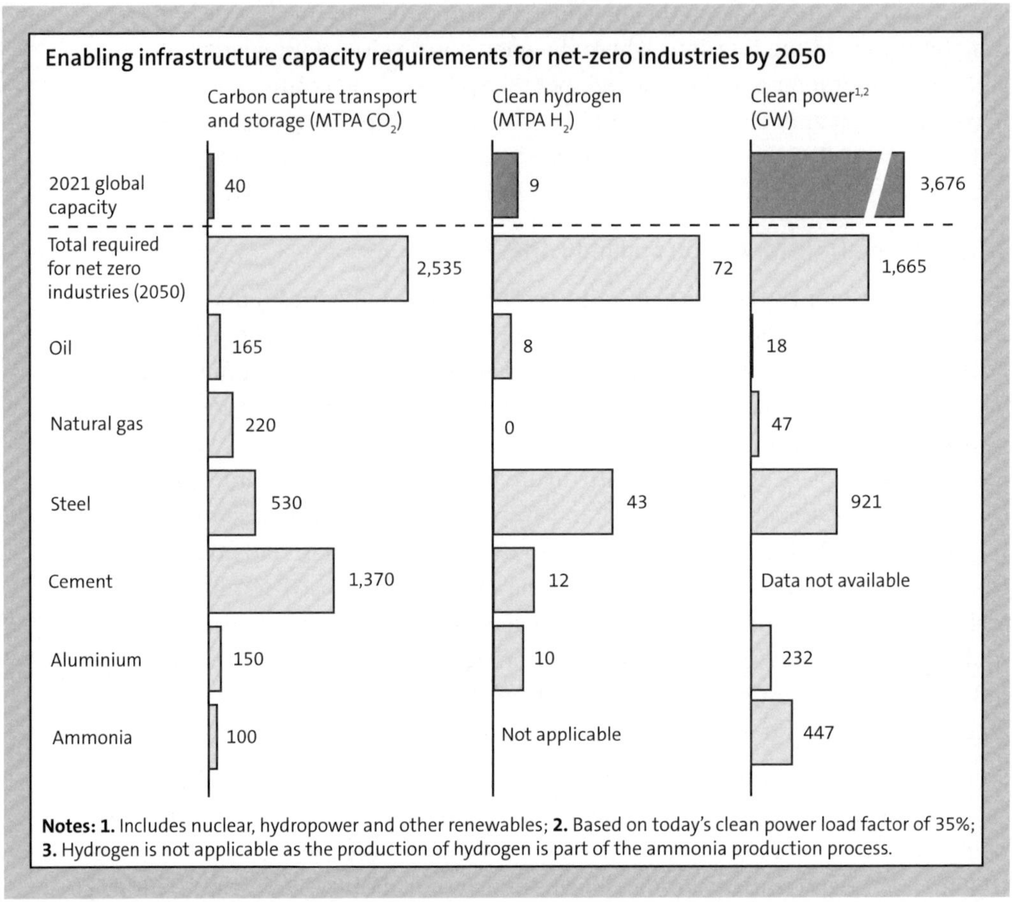

Enabling infrastructure capacity requirements for net-zero industries by 2050

	Carbon capture transport and storage (MTPA CO_2)	Clean hydrogen (MTPA H_2)	Clean power[1,2] (GW)
2021 global capacity	40	9	3,676
Total required for net zero industries (2050)	2,535	72	1,665
Oil	165	8	18
Natural gas	220	0	47
Steel	530	43	921
Cement	1,370	12	Data not available
Aluminium	150	10	232
Ammonia	100	Not applicable	447

Notes: 1. Includes nuclear, hydropower and other renewables; **2.** Based on today's clean power load factor of 35%; **3.** Hydrogen is not applicable as the production of hydrogen is part of the ammonia production process.

Figure 7.4 Carbon Capture, Hydrogen, and Clean Power Making Selected Industries Carbon Neutral (Source: World Economic Forum)

7.2.3 SAP's Role in the Transition to Sustainable Energy

Let's now talk about the tech and software industry in general and about SAP in particular. Of course, we run our data centers on clean energy, and SAP has committed to fully run on net-zero energy from 2023 onward—and to make our full value chain net zero by 2030. But the change for us goes beyond our own net-zero initiatives.

We claim that the world is running on SAP systems as much as it should be running on renewable energy: Our estimates suggest that about 80% of all business transactions worldwide touch at least one SAP system, so we must enable the business processes and business analytics that help our customers and the world economy transform to a system of renewable energy. On the global, national, and industry levels, the overall framework for this transition must be defined. But actions making a difference happen at the corporate, community, and personal levels. Koop focused on SAP's areas of action, as defined in the SAP Climate Strategy Framework:

- **Analysis: Getting the correct emission data (scope 1, 2, and 3)**
  Reporting and analyzing emissions across an enterprise's activities remain the foundation for all corporate climate protection strategies. Demands on data disclosure continue to increase with regulations requiring carbon content disclosures and with the consolidation of reporting frameworks through International Financial Reporting Standards (IFRS)-led global reporting standards.
- **Mitigation: Reducing emissions within the organization's span of control**
  Interventions in business-as-usual activities are needed to reduce emissions. To protect value (both financial and reputational), companies follow the greenhouse gas protocol[19] to reduce their own emissions (scope 1); switch to buying low-emission energy (scope 2); and purchase goods, materials, and services that come with a smaller greenhouse gas footprint (upstream scope 3). Interventions vary across the enterprise, but they are guided by a standards-based analysis of emissions factors.
- **Markets: Using carbon credits and markets to get to net zero and to buy and sell low-carbon products and services**
  Even ambitious emissions reduction targets will not eliminate all emissions. Net-zero targets will only be met by using the external carbon market to offset emissions that cannot be avoided or reduced. Low-carbon products and projects that create carbon sinks (like carbon capture and storage) will only create value by selling their credentials across enterprises in markets and networks. Carbon offsetting schemes require robust standards certificates to ensure their viability and trustworthiness in the carbon offset market.
- **Transition: Resetting business models as part of the energy transition**
  The global economy is under pressure to decarbonize by transitioning from hydrocarbons to carbon-free power generation, electrification, and the deployment of new energies at scale, especially biomass and and green hydrogen to transport and store energy. For some industries, simply managing emissions (actions across scopes 1, 2, and 3) isn't enough; the only way to grow in a carbon-constrained world is through transition and even reinvention.
- **Adaptation: Protecting your business from a changing climate**
  Resilience will be both financial (as the external cost of carbon is increasingly expensive) and physical. Physical resilience will be essential to avoid disruptions to operations and supply chains. The location of a company's assets and activities will be increasingly important to maintain business continuity, including insurance.

SAP's five pillars reflect mitigation and adaptation strategies. Mitigation must be considered at a global scale since greenhouse gas molecules stubbornly ignore borders, while adaptation is occurring at a local level. For example, many coastal cities are building defenses to protect against sea-level rise, and farming communities are moving to drought-resistant crops.

Koop also added this perspective:

> *We see this scramble for fossil fuels due to the shortages and peaking prices as a consequence of the war in Ukraine, and Putin using oil and gas as weapons. It is also a wakeup call for politicians and business leaders to look at the risks of being dependent on energy sources that are controlled by potentially unfriendly countries. Switching to renewable energy generated locally or by politically reliable partners becomes attractive, even if the cost per energy unit rises.*

For Koop, the carbon footprint of a barrel of oil heavily depends on its origins. Equinor in Norway enjoys short distances to their customers, and they use renewable shore power to run their oil rigs, resulting in low carbon dioxide emissions of 6.2 gram per mega joule (MJ) of energy produced (6.2 g CO_2/MJ). Producing oil in Saudi Arabia comes in at 7.9 g CO_2/MJ. Compare this rate to 30.5 g CO_2/MJ in Iran or 26.7 g CO_2/MJ for oil from tar sands in Canada.[20]

Passenger cars account for roughly 9% of global carbon dioxide emissions,[21] and even if many people still love their gas-guzzling pickup trucks or high-octane sports cars, passenger cars apparently can go electric and achieve zero (local) emissions comparatively easily. However, e-mobility requires a sprawling and expensive charging infrastructure, batteries are made from problematic materials, and drivers need to adapt to charging a car instead of refueling it. SAP is contributing to this transition by phasing out vehicles with internal combustion engines from the company car fleet of 27,000 vehicles. From 2025 onwards, our employees will only be able to order battery electric vehicles.

7.2.4 Decarbonizing the Energy System, One Molecule at a Time

Let's consider the perspective from David Rabley, Accenture's energy transition and sustainability global lead for energy, who has helped many clients navigate the transition to a low-carbon future. Rabley said that the journey to net zero needs to consider three fundamental elements that frame engagement with clients:

- **Understanding the system**
  "What is the role of today's energy system infrastructure, assets, capabilities in terms of moving to a lower carbon future?"
- **Accelerating the transition**
  "What is the set of actions that the entire energy system can take? Oil, gas, and utilities are the largest components of the energy system. The most obvious action is transforming the electrical energy system from predominantly fossil fuels to wind and solar resources."
- **Extending the frontier**
  "What are the emerging technology solutions and energy vectors that are going to

be important leading to a decarbonized future? In many cases today, they still need support (as in private and public financing) to achieve scalability and reliability and competitive cost per molecule or cost per joule."

Rabley said that, in the popular narrative, people love to jump to the last element—the frontier discussion—and to debate the role of synthetic fuels and green, blue, and pink hydrogen. He finds that this attitude jumps the gun, and he cautioned that no simple path exists to immediately decarbonizing the transportation, cement, or steel industries:

> *We first need to find solutions that work with today's assets. There is a significant set of near-term opportunities around driving efficiencies, driving increased energy management solutions, impacting demand, impacting the feedstocks of today's value chains.*

Rabley provided an example of the parasitic emissions of methane in the production of oil and natural gas. We've all seen the images of roaring gas flares over oil fields, but what you may not have seen is the methane leaking quietly from pipelines and gas fields and escaping in the atmosphere, where it is a potent greenhouse gas: Molecule for molecule, methane traps 84 times more heat as carbon dioxide over a 20-year lifecycle. Rabley pointed out that beyond gas that is flared or that goes into the atmosphere to do even more damage, methane is a product the chemicals industry buys as feedstock in production as well as a product that generates electricity and heats homes. The numbers are staggering: If we collected all the wasted methane, we could replace the natural gas that Russia had exported to Europe. According to Rabley:

> *What we're looking for now are opportunities for using data, using intelligence, using different approaches to find the right pathways to bring today's infrastructure forward, to monetize what we can. More than 50% of actions associated with methane can be delivered at positive economics for the industry globally, with an even higher percentage in North America where the Inflation Reduction Act will directly incentivize methane reduction from 2024 onward.*

Rabley sees potential for SAP to develop energy solutions that implement the value calculation for joint clients, integrating the relative impacts of short- and long-term revenues, costs, risks, and sustainability. In his client engagements, Rabley has found it difficult to piece this model together on a project basis and then turn it over to day-to-day operations.

On Rabley's point about understanding the energy system to decarbonize it, Beberness discussed the concept of "emissions intelligence" that SAP is piloting with Accenture:

> *If you look at what software companies are providing, including SAP, the focus is on outward-facing reporting, as with the SAP Sustainability Control Tower. But in addition, our CEO, Christian Klein, talks about a green ledger. For finance, companies have their credits and debits. They now require the ability to track their environmental, social, and governance credits and debits as well. SAP is the only company that can extend the financial ledger with a green ledger to integrate emission-related factors in daily operational decisions. Effective action to implement the net-zero strategy so many of our customers are committed to emission intelligence.*

Rabley and Beberness focused on emissions. We see an even broader role for extending systems for managing top lines and bottom lines through a "green line" that predicts and captures the impact of routine business decisions on the full set of sustainability metrics, including emissions, circularity, inequality, or land use. Integrating top line, bottom line, and green line enables employees on all levels of the enterprise to include sustainability and environmental, social, and governance (ESG) impact on their choices in the decision-making process, which is much more effective than seeing the result after the fact in a report.

Beberness provides an example from the oil and gas industry and his family life:

> *When companies are making decisions to build a carbon capture facility, they need to understand the ESG footprint of the facility equipment and the emission curve from running the facility at 100% or 50%. This is not so different from what we do when we go grocery shopping. My family buys organic food even though it's more expensive. We think it's good for the environment, it tastes better, and we don't want to put unnecessary chemicals into our bodies. In the long term, this will help us live healthier and longer lives.*

Implicitly, the Beberness family is running its own green line to complement the financial side of their grocery bill with harder-to-quantify but even more important metrics like taste, health, and longevity. Executives will find it harder to justify higher cost or lower profits with a positive impact of their decisions on "soft" factors such as environmental friendliness, higher recyclability, lower emissions, or fairer treatment of their supplier base. The green line we promote won't generate better stories around soft factors; it turns soft factors into hard factors. Beberness continued:

> *As companies are building their carbon capture, utilization, and storage facilities, they take into account not only the management of the equipment based on time of use, inspection data, lifetime expectancy, and so on, but also the lowest carbon dioxide footprint they can expect based on their maintenance strategy for the equipment. Emissions intelligence matters along the*

value chain because financial investors, regulators, and consumers will require it. One day when we fill our car with fuel or charge it, we'll not only see the price per gallon/liter or the charge per kilowatt hour but will also see the ESG number for that fuel or electricity. This will allow us to include these metrics in our purchasing and production decisions.

The SAP Emissions Intelligence program with Accenture identifies use cases where ESG information is relevant for daily business decisions to help deliver on net-zero commitments and to meet requirements from regulators, investors, and customers.

- **Monitor and measure**
  Linking carbon tracking across all transformations and transactions, with rich physical and business context data, weaves a continuous "carbon thread" along value chains.
- **Record, report, and audit**
  Automated reports that integrate ESG metrics with carbon controls and financial models like models from the Task Force on Climate-Related Financial Disclosures (TCFD) as well as create stakeholder trust.
- **Manage targets and performance**
  Setting, allocating, and tracking targets for science-based glide paths and across assets, business units, and employees, integrated with business performance and strategy metrics that shows how the organization progresses along its sustainability objectives.
- **Reduce, replace, optimize, and offset**
  Achieving the objectives requires defining, cataloguing, prioritizing, executing, and assessing actions. Incentives, in conjunction with other business objectives, can motivate interventions to reduce, replace, and offset emissions across all scopes.
- **Predict and rebalance the portfolio**
  Simulating the financial impacts of different emissions, carbon costs, and interventions at the asset level and the portfolio level helps anticipate the effectiveness and financial consequences of actions. This information can inform the definition of optimized capital plans, mergers and acquisitions, and operational and human resources decisions.
- **Trade and monetize**
  The differentiation of products and services—for example, through including ESG-attributed or compensated fuels, carbon offsetting transactions, and integrating sustainability metrics into equivalent annual cost (EAC) calculations—supports customers to better informed buying decisions.

These use cases support customer engagements, the identification of relevant business capabilities, and the design and required capabilities of new digital solutions from SAP and our ecosystem.

7.3 Oil and Gas

SAP's oil and gas customers have always dealt with the difficulties of running a business that is powered by big capital assets (i.e., pipelines, oil rigs and tankers, refineries, etc.) that have lifetimes of decades, while short- and long-term economic fluctuations and geopolitical, technological, or regulatory disruptions wreak havoc with the industry.

Beberness suggested that we imagine ourselves as executives in the industry, trying to run a profitable, sustainable, and responsible business:

> *Pre-COVID, oil and gas companies were starting to think about energy transition. They were driven in part by sustainability goals from the Paris Agreement. Then COVID hit. Travel ground to a halt, people worked from their home offices, production was reduced, so we saw a significant drop in demand for fuel. Oil tankers didn't move product; they were moored and used as floating oil reservoirs. In some parts of the world, the spot market price for crude oil was negative. Imagine that—you would get paid if you could syphon off some of the stuff.*

With emissions going down and the skies in China clearing up, sustainability-focused momentum from investors, customers, society, and regulation has increased. The combination of global warming and economic cooling have driven a renewed focus on sustainability.

Then, in February 2022, the Russian invasion of Ukraine shocked the world and the world market for oil and gas. Before the war, Russia's oil and gas made up 40% of the supply to EU countries. Sanctions throttled Russian supplies, and Europe scrambled to find new suppliers, creating new opportunities for oil and gas producers around the world. The urgency of wartime resulted in a production crunch with steep price increases and wild price oscillations: Every bomb dropped, every valve closed (or opened), every attack on pipeline infrastructure had the potential to upset markets, drive inflation, reduce industrial production, and even bankrupt companies.

Beberness commented on the perceived big opportunity for big oil:

> *You might think this would be a once-in-a-lifetime opportunity for the oil and gas industry. A competitor with a global market share of 10% gets isolated and there is a mad rush to replace him. Governments are suddenly requesting oil and gas companies to produce more to assist the EU. But this comes right at a point when companies are retiring their refineries that were no longer profitable, and they are retiring or laying off their workforce. Now try to turn this*

around and ramp up production, bring people back from retirement, and restart production with written-off assets. And you don't even know how long the demand (and the profitable prices) will last.

A month after the invasion, Houston, Texas, hosted the annual CERAWeek event dubbed the "Davos of Energy." Beberness recalled hearing a similar worry about the stability of demand for a product so unpredictably influenced by geopolitics. He said that industry leaders expressed a significant concern that, after increasing production, they'd be "hung out to dry" by customers if market conditions changed again.

The war in Ukraine also created public awareness for the easily overlooked dependency of our entire economy on hydrocarbon products other than energy. Some activists and politicians may think that it desirable to "get rid of all the oil and gas companies," but they may not realize how many products are made from oil and gas. Even experts may be surprised that bottling beer and making fizzy drinks can be severely impacted: High prices for natural gas result in reduced ammonia production for fertilizer, so there is also less carbon dioxide (a side product of ammonia production) to get beer into cans and bubbles into soda.

According to the Illinois Petroleum Resource Board (IPRB), one barrel (roughly 119 liters) of crude oil can produce either 39 polyester shirts, 750 pocket combs, 540 toothbrushes, 65 plastic dustpans, or 23 hula hoops—just a few examples among thousands of other products. You may argue that we could make do without pocket combs and hula hoops, but the IPRB also lists N-95 masks, cell phones, computers, and car tires—hydrocarbon-dependent products we would sorely miss. Essentially, pharmaceuticals and everything made from plastic or other high-performance materials depend on the oil industry. But Beberness has noted that awareness is still lacking:

> *Many people don't understand that if we shut off production or have a shortage of oil, they will not be able to buy lip balm, perfumes, adhesives, tires for their car, that cover for their smartphone, the smartphone itself, or even more critical needs like their prescription drugs. Our world runs on oil and gas. This is not something we can change overnight. We call it an 'energy transition' because it is just that: a transition.*

For Beberness, part of the challenge is how oil and gas companies communicate in the market:

> *It is interesting that we don't see and hear messaging from the energy industry on how important oil and gas is to our everyday lives beyond fuel for our cars. In conversations with energy companies, they say they don't believe educating would remove the stigma of the energy industry. I think they are*

missing the opportunity to explain how important the participation and partnership of the energy industry is to achieve the Paris Agreement targets.

Most people probably agree that we need prescription drugs, performance materials, and fertilizer. Thus, we can't just tear down the oil and gas industry—we need to rebuild it. This transition to sustainable energy requires massive research efforts and huge capital projects. The oil and gas industry has proven time and again that they are quite good at building and running capital assets over decades and in turbulent markets. According to Beberness, the industry has also been fantastically innovative in finding ways to match the growing demand for hydrocarbons with new production technologies:

> *If the world sets up the right regulatory and incentive framework for the sustainable energy transition, the energy industries will apply their brain and muscle to build scalable carbon capture and storage solutions, develop large-scale generation of renewable energy in remote regions of the planet, scale hydrogen technology to transport energy through space and time, and piece together synthetic hydrocarbons to make hula hoops and toothbrushes.*

Building, running, and commercializing all these technologies needs digital solutions, and for Beberness, SAP is well positioned to be a key player in providing the digital side of the transformation:

> *SAP is engaged with the energy industry through the sustainable energy transition. The solutions and capabilities that SAP provides to the organizations allow them to run their production operations and refineries today, and our solutions will also allow them to operate their new energy business tomorrow.*

He also found that the energy industry is already moving in the right direction, albeit with a specific focus in the different regions:

> *In the EU, I would say Shell, Equinor, Galp Energia, BP, and TotalEnergies have all taken very significant steps. TotalEnergies is the second-largest solar provider on the planet. They're moving into electric vehicle charging. Most of these companies have new energy businesses to drive their investments and resources towards their net-zero objectives.*
>
> *In North America, the oil and gas companies are more focused on carbon capture and biofuels, probably because of the significant domestic oil production that will get a new lease of life when carbon sequestration works at scale. The CEO of Oxy is very vocal about their carbon capture plans. They want to not*

only get themselves to net-zero but then sell excess carbon credits on the open market. That opens up a new market for them to sell to airlines and to manufacturing companies in hard-to-abate sectors.

With this model, the oil and gas company can offer to capture a predetermined volume of carbon dioxide on their behalf and would provide a certificate to the customer that can be audited and verified. Beberness noted that this carbon capture model would extend an oil and gas company's innovations beyond only meeting their own net-zero goals.

7.4 Utilities

Prompted in many cases by regulators, utilities have been transitioning to low-carbon electricity by making capital investments in solar parks and wind farms for longer than any other sector. We use "wind and solar" often as a term for technologies to capture "free" solar and wind energy, in the sense that the variable cost per energy unit is zero. However, wind and solar are both intermittent, so backup and/or storage is required to get us through dark nights, cloudy days, and lasting lulls. But profound differences between wind and solar energy become quite visible as soon as the technology leaves the lab.

7.4.1 Solar Energy

Solar energy has become popular with consumers in many countries. Solar panels on the roofs of many homes quietly and unobtrusively generate electricity, allowing the owners to show off their responsible lifestyles. Panels are a great conversation-starter with neighbors and other homeowners about yields, costs, subsidies, maintenance, unfair feed-in rates, net metering, rain and sunshine, and batteries. What makes solar energy also compelling is that the energy yield scales linearly with the size of the installation; in other words, a small installation on your roof has a similar efficiency, per square foot, as a football-field-sized solar park. Putting solar panels on roofs also means no additional land use, and hooking the panels up to the grid only requires a few short wires.

Utilities investing in large-scale solar farms and technology innovation have driven down the unit cost for solar panels and have increased panel efficiency, so even unsubsidized solar energy has become highly competitive.

However, solar energy is challenged by demand patterns and energy storage. Solar production increases as the sun rises to peak at midday, but that time is exactly when electricity demand from consumers goes down. Conversely, solar production goes down as the sun sets, which is exactly when electricity demand goes up again. Current power storage technologies limit how much solar energy can be "banked" during the day for use in the evening. The California Independent System Operator (CAISO) has identified a few opportunities ahead, such as improving battery and grid storage capacity with pumped-storage hydroelectricity and investing in peripheral renewable energy sources to fill in the gaps between solar supply and electricity demand.

Per Sellmann, a good example of this is currently being built in southern Germany: the world's biggest battery storage project, with a capacity of 250 MW. The idea of this grid booster project is to lower the cost of switching of energy creating assets like wind turbines when there is a surplus of energy in the grid, and to save it instead in those batteries; this would lower the redispatch costs (for security reasons) for utilities and end customers. Further, battery storage can perform a black start and helps to make the energy supply chain more resilient.

Solar energy from millions of residential roofs creates interesting challenges and opportunities for utility providers, which must ensure that energy generation and distribution infrastructure are stable and intelligent. Cost and revenue for energy supply and usage are allocated according to market dynamics and regulatory frameworks. People who own solar panel roofs have a much more intimate relationship to electrical energy than their neighbors who just plug in and pay the bill. After all, they are "prosumers" (both producers and consumers!) who spent time and money deciding on the right type and the right size of panels, figuring out financing and subsidies, completing the installation, and wading through paperwork before experiencing the big moment when, with the rising sun, the meter started to tally their homegrown kilowatt hours. The evolution from this is prosumers selling their surplus back to the grid, thereby becoming "flexumers." But energy flow requirements don't end there: Solar panels need insurance and maintenance, and the mismatch between supply during the day and demand in the evening doesn't go unnoticed, so why not install battery storage? Switching to an electric vehicle requires a charge point at home; how about a roaming scheme so you see the costs of charging your car on your utility bill wherever you go?

This will ultimately change the relationship between distribution grid operators and end customers, as the former gains more information about the end customers that traditionally resided only with the energy retailers. Utilities need to create a new customer intimacy and loyalty to move them along with their business transition. Even if you exclude solar roof owners from the persona definition, ideation workshops can generate a broad range of innovative services that utilities could offer to their customers. Banks used to say that customers who use more than three of their products are pretty much "unlosable," and it stands to reason that customers are less willing to switch providers for a marginally cheaper rate if they would also lose the value-added services they've booked on top.

7.4.2 Wind Energy

Wind turbines have come a long way from the antique and medieval windmills and the early days of wind turbines on Californian mountain ridges. Wind turbines have evolved into impressive machines with rotor diameters of up to 250 meters (820 feet). Some people find the sleek constructions beautiful, but severe pushback often arises against having one of these giants in your neighborhood. Some communities buy and operate wind turbines as a cooperative, so citizens can benefit from the revenue the windmill generates, which increases acceptance. The most effective wind farms are offshore where the winds are generally stronger and more stable. New technologies are arising that allow for floating wind turbines to be installed in deep water, allowing for further spread of the technology.

But of course, windmills share an annoying similarity with solar panels: In lulls and during storms, they can't produce electricity. Germans have coined the term *dunkelflaute* (dark lull) for time periods of more than one day when thick cloud cover renders solar panels useless, and a lack of wind brings the windmills to a halt. For these events, dispatchable power plants or energy stores must pick up the slack.

7.4.3 Renewable Energy and a Stable Grid

A power grid is stable if the energy that is put into the grid is equal to energy that is pulled from the grid. All power grids are alternating current (AC) grids with a frequency of 50 or 60 Hz and a wide range of voltages from hundreds of kilovolts (kV) to the domestic-use 230 volts of Europe and the 120 volts of the US and a few other countries. The power grid is essentially just a set of wires and transformers and can't store any energy. Sellmann explained what makes maintaining renewable infrastructure such a challenge:

> *In the United States, you have to keep the grid frequency at 60 Hz. If you drop too far below that frequency, say by loss of generation, it could trigger events in the power system causing voltage, frequency, and power imbalances, ultimately resulting in a system-wide collapse. So, you always need to feed as much power into the grid as you pull from it to maintain the 60 Hz.*

In Europe and other parts of the world, the grid frequency is 50 Hz but the challenge is the same. With fluctuating wind and solar generation, you typically have natural gas turbines or hydroelectric power as backups to ensure that sufficient power is fed into the grid to maintain the standard frequency. Pumped-storage hydroelectricity is essentially a battery: When you have a lot of wind and sunshine but not a lot of demand, the excess power is used to pump water uphill. When dark or not windy enough, these same pumps and motors double as turbines and generators. Pumped-storage hydroelectricity has been around since the 1890s and is a proven and reliable large-scale technology, but the number of appropriate locations for this technology is quite limited.[22] The industry is experimenting with other technologies to store electricity: compressed air in underground caverns, other gravity-based storage methods, flywheels, different kinds of batteries, thermal energy storage with molten salts, and of course using electrolysis to make hydrogen.

Sellmann argued that we need to solve the energy storage in the best case with solutions ready for black starts and quick on-demand availability of electricity to support renewables. Otherwise, we cannot use sun and wind to the extent that we want. We agree but also want to point out a second option: ensuring that energy demand is matched by supply in real time by adapting demand to the available supply. Initially, this approach sounds crazy: Would my utility provider really call me and tell me to dim the lights, stop cooking, and turn off my favorite TV show? On a large scale, this *demand-side management* has been around for many years: aluminum smelters, refrigerated warehouses, and other major electricity sinks are directly connected to utilities. And some of our customers already drive those demand programs within the US.

As Sellmann explained:

> *The early experience with renewable energy has motivated utilities companies to develop new solutions for their industrial customers. Now, the proliferation of storage batteries, electric vehicles, connected fridges and digitally controlled A/C systems creates new opportunities to engage with consumers and make them part of a decentralized 'virtual' power plant. If you allow your utility provider to remotely control storage batteries, your freezer, and when exactly your car gets charged, you probably won't even notice the demand side management, but you'll contribute to stabilizing the grid and maximizing renewable energy. But for this we require heavy investments from utilities in smart and intelligent technologies and infrastructure.*

Since the invasion of Ukraine, every citizen in Europe and beyond realizes how painful dependency on fossil fuels can be. Sellmann believes that this realization will accelerate the acceptance of new energy services and the development of new business models by startups as well as existing utilities:

> *In order to make the energy transition happen in countries with limited access to renewables, we need to involve not only industry customers like Volkswagen and Daimler, but also end customers. We also need to leverage distributed energy resources (DER) and decentralized energy generation to feed the surplus back into the grid. IDC predicts 10% of the worlds energy generation to come from DER by 2030.*

Sellmann's colleagues Markus Bechmann and Mateu Munar discussed new opportunities for utilities. Bechmann commented:

> *In the old energy world, a consumer was a point of delivery with a meter and an address for the bill. Now, utilities want to sell and deliver comprehensive energy services. The transition from consumer to prosumer is a big shift in the industry.*

Munar focused on an example from Australia:

> *We are starting to see 'community batteries,' where prosumers can store excess power from their solar panels. It is allowing utility companies to delay investments in substations. The good news for us: You need new digital capabilities to make these models work, and our SAP for Utilities solutions are the platform to develop, deploy, and operate these innovative services.*

Australia's Project EDGE[23] is blazing the trail. On the University of Melbourne website, the project claims:

> *It is the world's first project that brings together the spectrum of relevant stakeholders along the electricity value chain: customers, infrastructure owners, aggregators, distributors, the system/market operator, and researchers. Innovations will be demonstrated and tested in trials that will test operating envelopes and the trading of local services. This is crucial to understand the complexity, interactions, and challenges that distribution companies (and the electricity sector) will face globally as they continue to accommodate the widespread adoption of distributed energy resources.*

As summarized in Table 7.2, for Bechmann, traditional industry boundaries between utilities and oil and gas companies are blurring. Enel and Iberdrola are now piloting green hydrogen projects in southern Europe because there they have a lot of solar energy with little local demand. These companies want to convert electricity into green ammonia, which is easier to transport to industrial countries like Germany:

> *We see the rise of a whole new supply chain. Ammonia is a fuel and the raw material for fertilizer. Traditionally, ammonia is made from natural gas and going to green ammonia reduces our dependency on Russia and reduces carbon dioxide emissions.*

	Drivers	Challenges
Corporate sustainability	■ Regulation and corporate targets ■ Access to capital and investors ■ Public demand	■ Emissions calculation (for scope 1, 2, and 3) in corporate carbon accounting ■ Keep track of current position versus procured offsets with certificates for auditability
Grid-scale generation/storage	■ Demand for green energy from renewables ■ Volatile (partly negative) energy prices	■ Investment planning into renewables ■ Operations of generation fleet combining asset and market view
Transmission, distribution, and metering	■ Decentralized and renewable energy resources (DERs) ■ Demand for transportation capacity ■ Electrification of transportation	■ Enable grids to cope with DERs and high-volume electric vehicle charging ■ Avoid huge grid investments and improve flexibility ■ Support automated load shifting
Electrification and e-mobility	■ Switch to electric cars in many countries ■ Strategy shift to electric at many car manufacturers	■ Planning and rollout of charging infrastructure for e-mobility ■ Commercial settlement of e-mobility charging
Industry customers	■ Rising carbon dioxide prices ■ Demand for green hydrogen/ammonia	■ Decarbonization of entire supply chain ■ Prosumer engagement
Household customers	■ Customer demand for sustainability and prosumer services ■ Customer demand for omnichannel customer service	■ New business models for prosumer services ■ Operations of prosumer asset along the lifecycle ■ Holistic customer view ■ Support for automated load shifting

Table 7.1 Sustainability along Energy Utility Supply Chains

7.5 Hydrogen in the Energy Transition

Hydrogen is the simplest and most abundant element in the universe and stands out as the first element on the periodic table. It's also the most abundant element in your body: You need two hydrogen atoms for each water molecule (H_2O), and many more hydrogen atoms are required to form all the amino acids, fats, carbohydrates, and proteins we carry around with us. The "hydro" in "hydrocarbons" means that hydrogen (H) is also the prominent building block in methane (CH_4), diesel ($C_{12}H_{23}$), or gasoline (C_8H_{18}).

A molecule consisting of two hydrogen atoms (H_2) is a fantastic fuel: Throw in an oxygen atom (O) and a small spark, and hydrogen can lift rockets into space and leave just a cloud of steam behind. One kilogram (about 2 pounds) of hydrogen contains 33.3 kWh, the energy of about a gallon of gasoline.

Unfortunately, on Earth, pure hydrogen can't be found in the wild, so the other atoms that like to cling to hydrogen must be removed, which requires a range of energy-intensive processes, ultimately creating a rainbow of colorful hydrogen molecules:[24]

- Green hydrogen comes from electrolysis. Water molecules are split by electricity from renewables.
- Yellow hydrogen comes from electrolysis with solar energy.
- Pink hydrogen uses nuclear energy. The energy from splitting atoms is used to split water molecules.
- Blue hydrogen comes from reforming methane to hydrogen and carbon dioxide. In this process, the carbon dioxide is captured and stored.
- Turquoise hydrogen also comes from converting methane. The process delivers hydrogen and solid carbon, which is easy to dispose of or can be used as a raw material for a wide range of products.
- Gray hydrogen is also made by reforming methane, but the carbon dioxide is released into the atmosphere.
- Brown hydrogen is made from fossil fuels, primarily coal.

Obviously, making green, yellow, pink, blue, and turquoise hydrogen doesn't release carbon dioxide into the atmosphere and can be considered climate friendly. For our discussion, and for your future discussions with friends and colleagues, the rainbow of hydrogen illustrates the need to invest energy into making hydrogen. Since hydrogen is not a form of energy like solar or nuclear energy, a better approach is to think of hydrogen as an energy store or as feedstock for industrial processes.

SAP's Koop thinks climate-friendly hydrogen will be a "game changer" especially in industrial processes heavily reliant on fossil fuels today. In a trial in Sweden, a technique called *hydrogen breakthrough ironmaking technology* (HYBRIT) was used to make green steel.[25]

In consumer spaces, hydrogen competes with batteries and has tried to be the "winning fuel" in mobility settings. In the passenger car space, according to Koop, batteries have already won, even if companies like Toyota are investing in hydrogen passenger cars. In his opinion, "the race is over." In the transportation sector, technology and economics may favor hydrogen, but the jury is still out in this segment. Beberness added:

> *SAP participates in the CEO Alliance for Europe's Recovery, Reform, and Resilience. We are working with E.ON, ENEL, Iberdrola, and ABB to set up a cross-country green hydrogen value chain. The idea is to produce hydrogen where*

the renewable energy production is cheaper and to bring the molecules to the countries in Europe that have the highest demand. Increasing production of green hydrogen will reduce unit costs, and we will see even more hydrogen use cases.

A 2-gigawatt (GW) electrolysis facility can generate 40 tons of hydrogen per hour, about 100,000 tons per year. To put those numbers in perspective: The huge Fukushima power plant with six nuclear reactors produced 4.7 GW, so you would need 3 of the 6 nuclear reactors to create the 40 tons of hydrogen per hour.

Koop is particularly excited about green hydrogen from Australia. Fortescue Future Industries and its partners plan to produce 15 million tons of green hydrogen per year by 2030. For Koop:

“ *Australia today exports coal. To replace that revenue stream or even massively increase that, it's logical to invest in green hydrogen. In the future they could also become a big exporter of green steel because they have the iron ore. Green steel should be in high demand as cities grow with global population and increasing urbanization.*

Koop anticipated similar investment in hydrogen production in Middle Eastern and African countries, which enjoy a lot of space and sunshine like Saudi Arabia, as they look to replace oil revenues or develop new industries.

Given SAP's experiences in the oil and gas and utilities markets, Koop believes a major opportunity exists for SAP to support new and existing customers in developing their hydrogen business:

“ *In this space, we'll see startups and established energy players creating separate business units or joint ventures. They will need a flexible and scalable cloud ERP solution like SAP S/4HANA Cloud. The next phases are planning, financing, and building the new hydrogen facility, and SAP Enterprise Portfolio and Project Management can keep track of plans, activities, budgets, and risks. And then production starts, and our customers will need a new set of capabilities.*

In the production phase, SAP's intelligent asset management capabilities enable optimal operations and maintenance and its environment, health and safety portfolio support the safety of the operations, after all, hydrogen is highly flammable and explosive. In the supply chain, many oil and gas companies are using SAP Hydrocarbon Supply for primary and secondary distribution:

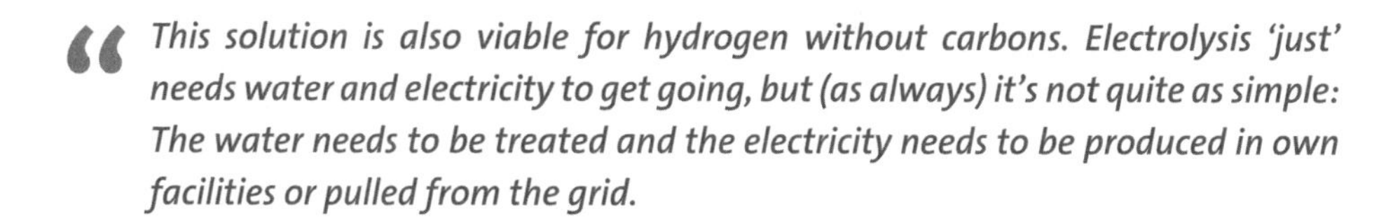

> *This solution is also viable for hydrogen without carbons. Electrolysis 'just' needs water and electricity to get going, but (as always) it's not quite as simple: The water needs to be treated and the electricity needs to be produced in own facilities or pulled from the grid.*

Electricity on the supply side and hydrogen as the product are commodities that are (and will be) traded on the market. As shown in the market turmoil caused by the Russian invasion of Ukraine, which has sent many companies into a tailspin, SAP Commodity Trading and Risk Management is a critical solution if contracts for buying supplies and selling products carry embedded price and volume risks.

Hydrogen production is also subject to scrutiny by customers, regulators, and investors regarding sustainability metrics, and in this area, the SAP solution portfolio can also come into play. For instance, GreenToken by SAP and SAP Product Footprint Management can calculate, track, and trace the greenhouse footprint of the rainbow-colored hydrogen molecules using the mass balancing method.

Accenture's Ashraf saw additional challenges for both the demand and supply side of hydrogen:

> *It's not necessarily the most efficient fuel to produce. It's not the easiest fuel to transport. And the economics from a demand standpoint in several sectors are still challenged, especially in the transportation sector where a lot of the focus has been.*
>
> *We forecast that the world will need 400 to 500 million tons of green hydrogen by 2050. If you look at where we are, despite all the talk about green hydrogen and the fact that electrolyzer technology has been around for a while, we only produce less than 1 million tons of green hydrogen. So you have to multiply today's production by 500 to 1000 times to really get to the level that is needed by 2050. That underlines the challenge with many of these 'frontier technologies.'*

Woods of ExxonMobil has commented on the hydrogen "color focus":

> *Europeans are making a strong case for green hydrogen made through electrolysis and using [renewable] electricity. Blue hydrogen is made primarily through methane, and then capturing the carbon dioxide that comes out. The US has a huge natural resource for blue hydrogen, we've got abundant methane, we've got abundant places to store carbon dioxide. And so, the answer for hydrogen in the US, I think, is going to be blue hydrogen. In Europe, they don't have the same abundance of methane, they don't have the same carbon dioxide capture capacity. So, electrolysis and green hydrogen makes more sense for them. And so rather than coming out with a dictate that says it's got*

to be green hydrogen or it's got to be blue hydrogen, let the market figure out how best to meet that.

Beberness agreed with Woods' position:

> *Before the war in Ukraine, green hydrogen was the focus; now blue hydrogen is also okay. Now we've integrated energy security into the sustainable energy transition formula. This adoption will allow for a much smoother and equally sustainable energy transition.*

Soon we may see the full hydrogen rainbow in the transition to green hydrogen. Understanding the sustainability footprint of the hydrogen mix, from the range of sources, requires tracking molecules along the supply chain from origination to storage to transportation to use. This requirement is similar to tracking the footprint of electrical energy that also flows through the grid, coming in varying quantities from a range of sources.

GreenToken by SAP provides the tools necessary to document a company's greenhouse gas footprint from their hydrogen-related activities in an immutable, blockchain-based shared ledger, as shown in Figure 7.5.

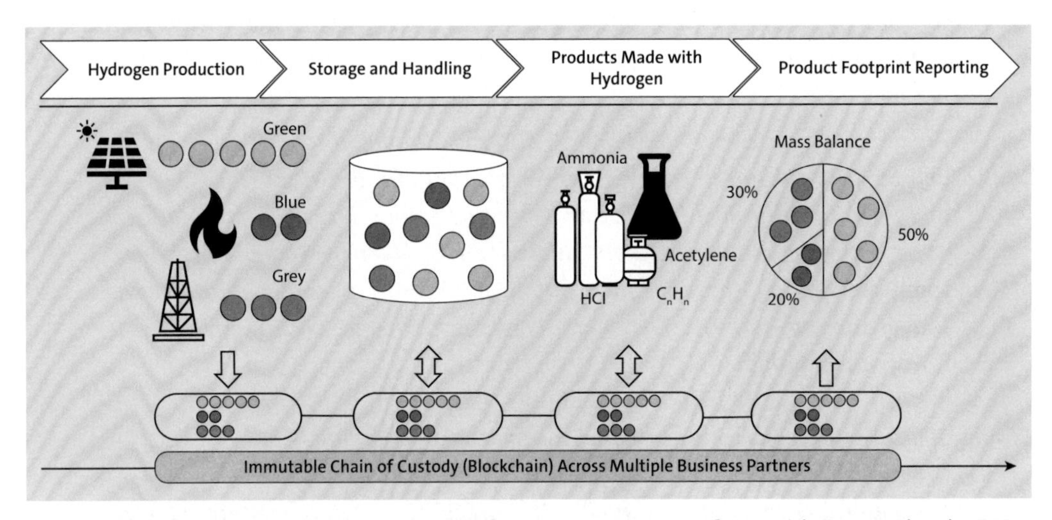

Figure 7.5 Tracking the Hydrogen Mix from Source to Point of Use with GreenToken by SAP (Source: SAP)

7.6 The Galp Energia Story

Galp Energia, SGPS, S.A., is a multinational energy corporation headquartered in Lisbon, Portugal. It traces its roots to 1846 when street lighting was first introduced in Lisbon. On its website, the company talks about its ambition and challenger spirit with which it has participated in the two largest discoveries of oil and natural gas of recent decades, in Brazil and Mozambique.

In recent years, Galp Energia has been expanding its retail business and investing in renewable energy sources. Beberness described the radical transformation journey Galp Energia has been on, led by a CEO who envisions the company becoming "the Amazon of energy." That CEO is Andy Brown, who arrived to Galp Energia in early 2021 after a 35-year career at Shell.

At CERAWeek in 2022, Brown outlined his goals for Galp Energia: to spend half its capital on zero-carbon dioxide sources of energy and to reduce its own emissions by 40% by 2030. To meet these ambitious targets, according to Brown, three things are key:

> *You need to change your portfolio. You do need to change your relationship with society and government. And you also need to change the way you lead your people.*

In an interview, Brown claimed a key role for oil and gas companies in the transition to renewables:

> *You have this emergent core of green solutions, particularly for the hard-to-abate sectors where oil and gas companies can fill the space. I think it's a legitimate space in which analysts expect there will be a future for oil companies.*

Galp Energia's core business is oil production in Brazil and Angola. The company has a refinery in Sines, Portugal, and it is a distributor of fuels and natural gas on the Iberian Peninsula. Beyond these existing products, Galp Energia has initiated a new generation of projects:

- **Renewable growth**
  Galp Energia operates power plants with an output of 1.2 gigawatts, mostly solar energy. It is developing renewables into a new core business by growing to 4 gigawatts by 2025 and 12 gigawatts by 2030 in Brazil, Spain, and Portugal.
- **Hydrogen and hydrogen-based fuels and biofuels**
  By the end of 2022, Galp Energia plans to launch a hydrogenated vegetable oil project. Waste from used cooking oil, animal fats, and biofuel is a product that can be used to produce diesel and aviation fuel. A next step is the production of ammonia as a fuel on the Iberian Peninsula or in Brazil. Galp Energia also wants to build a significant green hydrogen production with 600 megawatts electrolyzers to decarbonize its own refinery, which currently uses gray hydrogen from natural gas. Its plan is to gradually transform the traditional refinery at Sines into a green energy park.
- **The battery value chain and lithium hydroxide conversion**
  Europe will require something like a ten-fold increase in battery capacity by 2030.

Currently, all raw materials are imported. Galp Energia wants to create battery-grade lithium hydroxide. Portugal happens to have the largest lithium resources in Europe, the eighth largest in the world. Galp Energia is investing in a joint venture with Northvolt, one of the largest independent battery manufacturers in Europe, to build a lithium hydroxide conversion facility.

- **Carbon capture**
  Galp Energia is pursuing CCS for the carbon dioxide emissions at Sines. Today, Galp Energia doesn't have distinctive operating capabilities, unlike other big oil companies. It plans its own project as a foundation for conducting CCS projects elsewhere.

Brown then discussed the role of Galp Energia's SAP systems in the organization's transition:

> *We have recently rolled out SAP S/4HANA. I think we're the first international oil company to do a full rollout. I think it's been extraordinarily well executed. My experience with new systems is that the teething problems can be quite painful. I can't say our rollout has been entirely flawless, but it's been much smoother than I have expected.*

Brown said that an effective ERP system is critical to Galp Energia's plan to create completely new businesses in new geographic regions:

> *If you don't have a good, central enterprise architecture, then you know you're not going to have the control foundations that you need to manage those new ventures. An up-to-date enterprise architecture is crucial, and I think SAP has been a good partner for that. But SAP also is a leader in embracing the energy transition, so I think it has been very supportive of the ambitions that we have.*

He even hoped that the oil and gas sector would become more ambitious and saw creating credibility in sustainable energy investments as a central issue for oil and gas producers:

> *How we build credibility around investing large amounts of capital in the new energy space is, I think, is a crucial thing for oil and gas companies to come to terms with. This middle ground is biofuels, hydrogen, and carbon capture and storage because it's business we understand, we can demonstrate it, we can deliver in that space. But we have to place very big bets to create momentum for the energy transition. It will need a lot of commitment, dedication, investment, risk-taking on new technologies and new value streams that the companies have yet to demonstrate.*

As Beberness described earlier, oil and gas companies have unmatched capabilities to make big, risky bets—and Brown brought an "If not us, then who?" mindset, encouraging companies to shoulder leadership roles in this energy transition.

7.7 The Shell Story

You may find it hard to believe: The global energy giant Shell had its beginnings in a London antique store in 1833. One day, Marcus Samuel decided to expand his offerings with seashells, which were popular with interior designers at the time. The product took off, and his trade boomed. What was once an import-export business evolved over time, just like its iconic logo, which became one of the most recognized around the world. Now a global enterprise operating in over 70 countries, Shell reported revenues of $261 billion in 2021, ranking fifteenth in the Fortune 500.

Shell has a proud history of innovation. In the 1890s, it pioneered safe shipping through the Suez Canal and rolled out new aviation fuels in the 1930s.[26] Today, Shell is pioneering next-generation liquefied natural gas (LNG) carriers and fiber optic sensors in wells. Shell's history archive extends over miles of shelves in its office in the Netherlands and in museums around the world.

A few years from now, the archives will document the massive transformation Shell has undertaken with its "Powering Progress" strategy, which lays out the company's plans to accelerate the transition of its business to net-zero emissions.[27] Four guiding principles frame this strategy: generating shareholder value, powering lives, achieving net-zero emissions, and respecting nature.

At an investor presentation in 2021, then-CEO Ben van Beurden outlined Shell's target to become a net-zero emissions energy business by 2050. This goal is ambitious because it includes scope 3 emissions (in other words, emissions from customer use of Shell's energy products), which account for over 90% of its total emissions. Shell sells around three times more oil and gas products than it extracts and refines oil or gas.

There is an ongoing discussion about accountability for emissions from producing and using fossil fuel products. Most energy players see their responsibility to become net-zero for their own scope 1 emissions, scope 2 emissions, and inbound or "upstream" scope 3 emissions, which would include emissions from building capital assets, from consumables, and from all the other resources used for making and selling fossil fuel-based products. We think this is a perfectly reasonable position because it addresses the emissions that can be influenced by the producers of hydrocarbon products. But there is also a discussion to be had about the outbound or "downstream" scope 3 emissions. Some people want to hold the energy industry accountable for the emissions produced when *using* their products. Let's be very clear about what this means: The 2.3 kg (5 lb) of carbon dioxide emissions that will sputter out of the tailpipe of your vintage car in 2050 would have to be compensated by the industry to no longer show up on your emission balance. This discussion is of course not settled, but we believe that people who use a product—for example, to participate in a Sunday morning classic car rally—should foot the emissions bill, not the producers of the fuel.

Over the course of our long relationships with key industry players, we have seen how business models are transforming. Traditionally, the focus was on securing and developing assets—a business model linked to commodity prices and project-based returns. In recent years, we have seen new approaches to creating value from advanced products that provide low or no carbon energy and from risk management and related services. A lot of capital investment is flowing in this direction, making cash flows less exposed to oil and gas prices and more connected to the broader economic growth.

Since the 1970s, around the time of the OPEC oil embargo, Shell has developed "scenarios" as a way of helping governments, academia, and businesses explore the possibilities and uncertainties ahead in the energy system.[28] The future may not be predictable in detail, but it's also not completely random. A proven approach to support business decision-makers is the development of scenarios that outline possible pathways into the future and articulate their implications for the business and its stakeholders.

Shell's energy transformation scenarios explored three possible future pathways: the "Waves" scenario emphasizes wealth, the "Islands" scenario emphasizes security, and the "Sky 1.5" scenario focuses on health.[29] In all three scenarios, the energy system is decarbonized, but the speed at which this happens differs considerably. While all three scenarios share common milestones, those will be reached along different timelines.

All assume substantial improvements in energy efficiency while the overall global energy consumption continues to grow. They all involve significant electrification of the global economy, with renewable resources dominating. In sectors that are harder to electrify (like aviation or ocean shipping), demand for energy-dense and portable liquid and gaseous fuels will continue. However, all scenarios also have progressively deep decarbonization in these sectors as biofuels and hydrogen-based energy options gain ground.

Finally, all three scenarios rely on an increasing level of emissions removal with technology and by nature. As a result of the different paces of decarbonization in each scenario, they all achieve net-zero emissions at different times in the future. This difference directly impacts the average global temperature rise above preindustrial levels for each scenario.

Carlos Maurer, executive vice president for sectors and decarbonization, laid out Shell's strategy for decarbonizing global sectors, using aviation as an example:

> *It includes participating in the Clean Skies for Tomorrow initiative for aviation and our effort around Sustainable Aviation Fuel (SAF). It currently accounts for less than 0.1% of the world's consumption of aviation fuel. We announced our*

ambition to have at least 10% of our global aviation fuel sales as SAF by 2030. We are working with Rolls-Royce to test 100% SAF in airplane engines for the first time.[30]

As summarized in Figure 7.6, Shell's Energy Transition Progress Report for 2021 listed the variety of energy pathways for some sectors, noting that a fifth supply—mitigation and offsets—focuses on CCS and on nature-based solutions and is applicable across all sectors.[31]

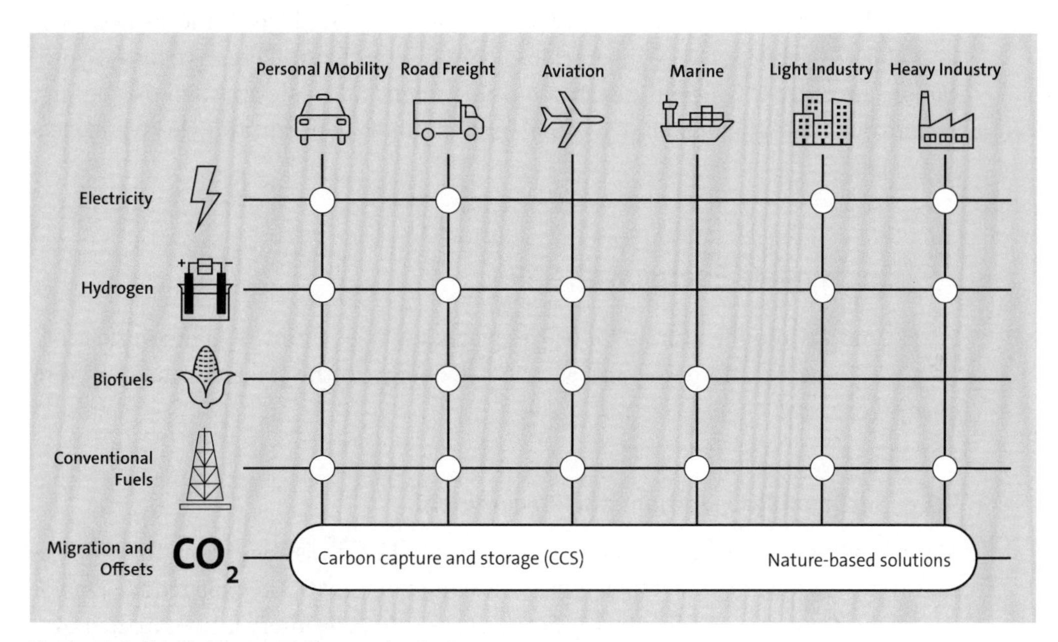

Figure 7.6 Shell's Energy Pathways by Sector

Shell and SAP agree that the first step to decarbonize the economy is a standards-based analysis of the current sustainability footprint of their own operations and the overall value chain. Shell is a founding member of the Open Footprint Forum and SAP is working with Shell to create a common model for footprint-related data covering all emission types, consumption, and base calculations to normalize and aggregate industry data.[32]

But we expect to take concrete steps to adopt Shell's Accelerate to Zero fleet solutions as it rolls out its own electric vehicle charging infrastructure.[33] We also comment on this initiative in a feature in SAP's News Center.[34] Shell is a big player in the energy industry and has significant influence on its peers in the industry and its customer base across a broad range of industries. While Shell is a "powerhouse" in energy, we consider SAP to be a powerhouse in the digital space to provide the digital solutions that enable the insights and actions to move the energy industries into a sustainable future.

7.8 One Goal, Many Pathways to Energy Transition

Accenture's Ashraf emphasized the need for flexibility during the energy transition, which extends to several industries and geographic regions at different paces. For example, Accenture is collaborating with Equinor, Shell, and Total on a carbon capture and utilization storage project called the Northern Lights. Meanwhile, most industry leaders are developing a broad portfolio to develop solutions during this energy transition.

Time will tell which initiatives and ideas will develop the biggest momentum and impact in the transition to a sustainable energy system that will preserve fossil hydrocarbons for more intelligent uses than burning them and create a system that doesn't require our great-grandchildren to pedal for 10 hours to generate a single kWh of electrical energy.

7.9 Sustainable Energy with SAP

The oil and gas and the utilities industries play a key role in providing the world economy with reliable, affordable, and sustainable energy in the transition to a net-zero energy economy.

7.9.1 Sustainable Operations and Procurement

The energy industries need to significantly curtail their emissions by up to 90% relative to current levels to meet their emissions reductions goals. They can reduce a major part of their emissions by first taking the most cost-effective interventions in their own operations. Support from supply chain and procurement in combination with sustainable operations can reduce scope 1 and scope 2, and upstream scope 3 (for purchased goods and services and inbound transportation) emissions, contributing to energy products with low-to-no greenhouse gas emissions. SAP solutions support this transition:

- SAP Intelligent Asset Management supports reliable and predictable operations to improve sustainable outcomes.
- The SAP S/4HANA Cloud for EHS solutions for environment management, health and safety management, and incident management identify, analyze, and mitigate environment, health, and safety risks. This includes safely handling and storing chemicals, monitoring industrial hygiene, reducing environmental impacts, and supporting compliance with emission-related regulations.
- GreenToken by SAP supports tracking and tracing greenhouse gas product footprints in an immutable shared ledger.

- SAP Ariba Supplier Risk and SAP Ariba Supplier Management support responsible sourcing from suppliers with risk profiles to ensure business continuity and avoid supply chain disruption.
- SAP Business Network for Logistics provides material traceability along supply chains to identify and mitigate potential disruption.

7.9.2 Carbon and Emissions Intelligence

Avoiding, reducing, and compensating for carbon dioxide and other greenhouse gas emissions is a complex endeavor. Standards-based calculation and accounting for direct and indirect emissions along the value chain are prerequisites to documenting regulatory compliance and to identifying and prioritizing reduction initiatives. These include carbon capture and storage and compensation for residual emissions with carbon credits, tokens, and offsets. SAP solutions provide insight and support climate action:

- SAP Intelligent Asset Management supports reliable and predictable asset operations.
- The SAP S/4HANA Cloud for EHS solutions for environment management and incident management identify, analyze, and mitigate environment risks. This contributes to reducing environmental risks and to ensuring compliance with emission-related regulations.
- SAP Treasury and Risk Management provides comprehensive functions to secure against financial risks with accurate registration and a comprehensive view of all business activities.
- SAP Commodity Procurement, SAP Commodity Sales, and SAP Commodity Risk Management support the commodity end-to-end process from planning, buying, and selling with management of price risks, and hedge accounting.
- GreenToken by SAP supports tracking and tracing of the carbon intensity of energy products including the certificates of origin for renewable energy production.
- SAP Commodity Management, Option for Deal Capture supports renewable identification numbers (RINs) used by the Renewable Fuel Standard (RFS) program in the US to keep track of renewable fuels.
- SAP Upstream Hydrocarbon Accounting and Management keeps track of volumes, revenues, and cost of all produced hydrocarbons.
- SAP S/4HANA Oil and Gas for hydrocarbon supply and primary distribution supports the end-to-end hydrocarbon logistics processes.

7.9.3 Low Carbon-Intensity Fuels

The hydrocarbon supply chain of the future needs to provide affordable, renewable, and low-emission energy resources. Applying sustainability and circularity principles lead to a resilient and climate-friendly fuel system with a mix of liquefied natural gas (LNG), hydrogen as an energy store, biofuels, and synthetic fuels with high-energy densities for aviation and transportation. The following solutions contribute to a low-carbon fuel mix:

- SAP Enterprise Portfolio and Project Management helps manage risks, activities, budgets, and costs to successfully deliver low-carbon projects and programs within their budget, time, and outcome constraints.
- SAP Intelligent Asset Management helps to reliably and efficiently operate energy generation and distribution assets.
- SAP S/4HANA Cloud for EHS environment management identifies, analyzes, and mitigates environmental risks of operating a supply chain of renewable and low-emission fuels.
- SAP S/4HANA Oil and Gas for hydrocarbon supply and primary distribution handles end-to-end logistics of low-carbon intensive fuels.
- GreenToken by SAP supports the mass balancing, tracking, and tracing of greenhouse gas product footprints in an immutable shared ledger.

7.9.4 Renewable Electrical Energy

Renewable generation additions are a global priority driven by the need to reduce greenhouse gas emissions. In the EU alone, the current plans are going from 22% of renewable electricity in 2020 to 45% in 2030.[35] This transition requires massive capital projects to create the infrastructure and business practices to efficiently operate distributed energy resources, supported by SAP solutions:

- SAP Enterprise Portfolio and Project Management can be used for planning and delivering large-scale energy infrastructure projects.
- SAP S/4HANA facilitates integrated enterprise asset management of distributed energy infrastructure.
- SAP Intelligent Asset Management offers integrated anomaly and damage detection in the operation of renewable energy infrastructure.
- SAP Analytics Cloud supports integrated reporting and analytics requirements.
- SAP Utilities Core supports integrated billing of renewable energy generated and stored by a complex system of solar, wind, and battery resources.

SAP, German utility STEAG, and research institute Fraunhofer-Gesellschaft work on an innovative cloud solution for integrated energy planning and optimization.

7.9.5 Industrial Energy Management

An energy system with a growing share of renewable electricity requires close coordination between supply and demand. We can no longer allow supply to satisfy the anticipated demand by simply shoveling more coal under a boiler. Instead, industry needs to integrate the anticipated supply of wind and solar energy into their production planning to minimize the strain on the grid and energy storage systems—and to benefit from low energy prices on sunny and windy days and in times of low demand from households. SAP solutions can support industry in the transition to adapting electricity demand to the available supply:

- SAP Integrated Business Planning can factor energy supplies and energy costs into overall business planning to create cost-effective and robust operations.
- SAP Intelligent Asset Management supports reliable and predictable operations aligned with the energy supply-driven operations.
- SAP S/4HANA Cloud for EHS environment management handles emissions resulting from energy production and consumption and runs a consistent process to support the full lifecycle of compliance-driven tasks and exceptions.
- Flexinergy, a solution by SAP partner Evolution Energy, provides integrated energy and environment monitoring that collects energy meter data in real time to support energy billing, front office, and risk operations.
- SAP Manufacturing Integration and Intelligence supplies energy monitoring and analytics functionality to collect energy information and analyze energy consumption from production assets.
- SAP Business Technology Platform provides data management capabilities to acquire, store, and analyze energy data and integrate it with business processes.
- GreenToken by SAP documents and communicates energy sources and related emissions information along the supply chain.

7.9.6 Evolution of the Digital Retail Service Station

Charging the battery of an electric vehicle takes longer than topping up a fuel tank. This creates new opportunities for service stations that have captive customers for something like 20 or 30 minutes, even if their cars are plugged into fast chargers. And it may get even better; imagine a virtual service station that integrates the driver's home, workplace, or vacation destination! Incentivize a few college graduates in a design thinking lab and you'll be amazed by the range of suggestions they generate in a few hours. Ideas may be easy to come by, but turning them into reality needs a handful of SAP solutions:

- SAP S/4HANA Oil and Gas for secondary distribution management automates and controls all sales and logistic processes of fuel products.

- SAP S/4HANA Oil and Gas for retail fuel network operations efficiently manages a service station network to maximize customer satisfaction and loyalty.
- SAP S/4HANA supports convenience retailing, sales, and marketing with capabilities for assortment management, promotion management, continuous station replenishment, and settlement management as a foundation to grow the customer's basket size and improve their experience.
- SAP E-Mobility optimizes electric vehicle charging and billing services for drivers at service stations, at home, at work, and in public locations.

7.9.7 New Customer Engagement Models

The energy transformation enables a range of new business models for utilities. Accurate metering and correct billing and invoicing are table stakes. A new focus on a sustainable customer relationship and customer lifetime value can define and capture new opportunities to maximize customer satisfaction and increase the share of their energy and convenience wallet. SAP solutions support the transformation from a meter-centric to a customer-centric business:

- SAP Customer Experience facilitates a personalized, multi-channel experience of marketing, commerce, and service to capture business with the new generation of "prosumers."
- SAP Utilities Core provides a complete cloud solution for energy utilities with built-in intelligent technologies, including AI, machine learning, and advanced analytics for more efficiency and better customer service.
- SAP S/4HANA Cloud for receivables management improves the speed and accuracy of accounts receivable with a unique integrated suite of intelligent finance applications.
- SAP Enterprise Asset Management solutions improve asset performance and reliability.
- SAP Cloud for Energy offers integrated management of smart meter data, including data validation, estimation, and editing.

Chapter 8
Circular Economy

The Mythology of the Circle

In the 2016 science fiction movie *Arrival,* linguist Dr. Louise Banks is recruited to help understand seven-legged aliens who communicate with inky circles. Once decoded, the circles communicate their cyclical worldview where the past, present, and future can all be seen at the same time, bending the linear flow of time, upsetting the principles of causality and breaking Ludwig Boltzmann's unbreakable laws of thermodynamics.[1]

Arrival stands in the long history of mankind's fascination with circles. Artist Thoth Adan has written, "As Pythagoras would say, the circle is the most perfect shape. It withholds all, and everything emerges out of it. Every point of the circumference is the exact same distance from the center. The circle has no beginning and no end; it is infinite and stands for both non-existence and eternity."[2]

Circles show up in many spiritual settings. In Christianity, three interconnected circles represent the Holy Trinity. In Islam, the circle is a symbol of the sky and infinity. *Mandala* is Sanskrit for "circle" and plays a major role in Hindu and Buddhist culture. The *ouroboros*—the snake eating its own tail—comes from Egyptian mythology.

Circles are also pressed into service for less spiritual purposes: three interconnected circles in the Krupp logo symbolize their patented seamless railway wheels, four overlapping circles adorn Audi cars, five rings in the Olympic symbol stand for the five participating continents.

Modern humanity has institutionalized circular thinking with the preface "re-." Terms like return, recycle, refurbish, and re-commerce are increasingly common in today's digital commerce. But they have been a long time in the making. The Japanese have been reusing their paper since the eleventh century. We don't know how many treasures we have lost when monks scraped parchment clean for reuse[3] and how many pieces of art get lost when files are deleted to recover digital storage space. Not even monarchs are exempt from the reuse cycle. In 1776, during the War of Independence, a statue of King George III in New York City was torn down, melted, and converted into bullets.[4]

In some ways, humanity has made significant progress. Aluminum recycling is a success story: Nearly 75% of all aluminum ever produced is still in use today.[5] Recycling aluminum only requires around 5% of the energy needed to mine the bauxite and make new aluminum. A similar percentage is enjoyed by Porsche, which claims that two-thirds of all Porsches ever built are still on the road or in museums.[6] The steel and aluminum of scrapped Porsches can live on as shipping containers or as beer cans. Who knows, maybe a few atoms of James Dean's iconic Porsche 550 Spyder sit in your fridge right now!

Beyond recycling, we should also be thinking about reuse, repair, redeployment, and many other circular techniques. It's time to rethink our throwaway, single-use lifestyle.

8.1 Linear Societies

You may be shocked that, after millennia of our obsession with circles, only 9% of the 100 billion tons of material produced globally is recycled every year. Per Stephen Jamieson, global head of circular economy solutions at SAP:

> *This essentially means 91 billion tons a year of material is being left to a fate in the natural environment every year. And we continue to extract another 91 billion tons of materials every year to produce brand new versions of so much.*

Even the success with aluminum leaves plenty more to do: the US recycles only 55% of its aluminum, whereas Brazil leads the world with 97%.[7]

Jamieson likes to use the expression "dark side of the moon" to explain how little we have focused on how much "stuff" we produce, consume, and waste on our small planet. "The challenge with the linear economy today," he said, "is that the moment that something is sold, the lights go out and there is little visibility on the whole flow from point of sale through to plastic trash floating in the ocean." He cited a few key statistics to illustrate that the "take-make-waste" economy is at a tipping point: 90% of biodiversity loss is due to the extraction and processing of raw materials and 9.3 million pieces of microplastics are found in each cubic meter of ocean water (35 thousand in each gallon).

Jamieson explained that global poverty and an overwhelmed planet play critical roles:

> *We know that nearly two-thirds of the world's population don't have access to a decent waste management infrastructure. On one hand, you have the industry looking to massively increase its output. On the other hand, we see a society that is completely ill-equipped to do anything with that output after its end of useful life other than burn it or leave it to a fate in the natural environment. That's the key challenge.*
>
> *At the end of the day, these materials are not being produced to deliberately harm the environment. They're being produced because there's a demand for the products.*

According to Jamieson, customers aware of sustainability issues can initiate change by moving their demand to sustainable products:

> *So, it may start with a 'What can I do for you today, dear customer?' kind of conversation. 'This pair of jeans looks lovely, but I wonder just how many hectares of land have been used to grow the cotton and how many cubic meters of water have been used for irrigation. What happens to the land? Where did the water come from and where did it go? Is the environment in a better or worse state after this pair of jeans has been made? Where's the evidence?'*
>
> *These are perfectly reasonable questions for the consumer to ask, and these are exactly the questions that we need to help our customers answer with our sustainability solutions.*

Jamieson said that this pressure, in turn, has prompted measures from government, industry, and non-governmental organizations (NGOs):

> *The opportunity here is to aggregate insights from multiple customers and then work with the NGOs, policymakers, and investors. That allows manufacturers to understand the market demand for different materials and their environmental impact in the target markets. This informs more predictable investments and better policy measures that actually accelerate a systemic solution. In the context of a circular economy, we're generally talking about extended producer responsibility, including plastic taxes and other measures.*

We have all seen images of plastic bottles floating in the oceans and plastic waste burning in landfills. Now imagine if lemonade in plastic bottles could only be sold in places with an infrastructure to process (preferably reuse or recycle) the empty bottles, or if you could also ban the "export" of plastic waste from Europe to Africa. We think it is safe to assume that in this scenario, industry and governments would get all creative and find ways to be compliant with these rules and still sell their products. Maybe they would invent environmentally friendly, reusable containers? Implement a deposit system that makes the bottles come back to the retailers and manufacturers? Onshore processing plants in Europe that keep materials in the loop?

Jamieson noted that institutions are betting on sustainability as an investment driver:

> *The investment community takes action. In the period from 2016 to 2020 we've seen a 10-fold increase in the number of funds that focus on investments that support a circular economy.*

8.2 Thinking Deep and Wide about Material Flows

Circularity is not a concept that equally applies to all industries and all materials. We believe the most effective approach to build circularity into our economy starts with the end-to-end material flow that can cross industries and end up in surprising places. If you grow wheat, you'll end up with straw as waste. The straw will decompose on the field mostly into water and carbon dioxide—circle closed within a year. But you can try to expand and slow down the circle by looking at additional uses. For example, straw is a good insulator, so you can make a building material from it that will last for many decades until the house is torn and the straw goes to landfill—or back to the field as organic fertilizer. Discovering uses for materials that are waste in one industry but a valuable input in another industry is probably the best way to cycle materials slowly and through as many stages as possible.

8.2.1 A Vertical Approach

Jamieson said he believes we need to take a "vertical" approach (i.e., by industry and subindustry) to the type of major materials we use:

> *Each material has its own role in the circular economy and has its own challenges and dysfunctions. And, as you'd expect, each material needs to be treated with its own set of solutions.*
>
> *Take textiles: To produce cotton clothes, we use 250 billion tons of water annually. Yet, 85% of that fabric goes into landfills after only five or six wears."*

At least cotton decomposes in landfill, other than clothes that are made from synthetic fibers that can last for centuries. SAP has identified and catalogued other materials—plastic, food, building materials, electronic components, and batteries—that create a significant waste impact and an opportunity for SAP to work with customers on intelligent solutions for more circularity.

> *Building materials account for 50% of total resource consumption and 39% of world energy emissions. That includes cement manufacturing and transportation of heavy construction materials. In the electronic components space, we're losing $57 billion a year in the form of e-waste.*
>
> *Or look at food. The vast majority of the agricultural land that we have in this world is used to grow four species of crop: wheat, rice, corn and soybean. That's utterly crazy, if you think about it. There are thousands of species that we could be sourcing our food supply from. There's a biodiversity opportunity, and a natural capital opportunity from having a rich and varied agricultural system.*

8.2.2 Circularity for Food and Farming

Admittedly, not all solutions for difficult challenges are digital. As Jamieson noted, some tap into centuries-old farming practices, and into a decades-old understanding of biology and chemistry.

> *If you plant peas next to barley you can start natural nutrient cycles: peas fix nitrogen so you need less nitrogen fertilizer that is made from natural gas.*[8] *Each individual geography on Earth has an optimal mix of plant life. If we were able to adapt the diet of people based on the geography, we would create massive benefits to the global ecosystem.*

Anja Strothkämper, vice president for agribusiness and commodity management at SAP, agreed with the possibilities of more sustainable agriculture.

> *Crop rotation and variation is indeed a way of being more natural about the soil rather than applying tons of fertilizer or crop protection. The food industry knows that the same crop tastes the same and the consumers like the acquired taste. But I think the industry is getting more flexible and sees an opportunity with customers who don't fancy crop protection chemicals. If we buy locally and accept or even seek variation in taste, then the industry can offer a compelling variety of fruits and vegetables. I think this is all changing. Paradigms are changing.*

Strothkämper talked about software support for farming, particularly about the SAP S/4HANA suite. "We have a farm management solution for larger enterprises," she said. "A few years ago, we took first steps into the farming world to learn what is needed and how the industry works. We asked customers, 'What are you planting? How are you irrigating? What fertilizer and crop protection do you use?'"

SAP solutions such as SAP Intelligent Agriculture are enabling data-driven farming operations and services. This is a fully cloud-based solution, natively developed and designed to model micro-data and big data on the field. We take data from weather satellites and weather stations, drone and satellite images, and soil sensors to understand what's happening in the field. The data is used to train machine learning models that optimize farming. SAP Intelligent Agriculture covers four focus areas:

- Farm data management allows disparate farming segments to manage their data consistently across the enterprise using a broad, crop-independent model.
- Farming processes include task management, field planning and history, and map-based overviews.

- Farming intelligence provides intelligent planting recommendations. The solution uses structured and unstructured data for data science scenarios and creates or integrates decision support models.
- Integration of partner solutions creates an innovative ecosystem for specialized solutions from startups, from manufacturers of fertilizer and crop protection, from providers of farm equipment.

Strothkämper summarized the connection between the industry and the environment:

When you look into agriculture and climate change, yes, this is an industry that contributes to more than 30% of greenhouse gas emissions, so that's a lot.

The interesting part is that you have emissions coming from fuel and the fertilizer production. You have emissions from deforestation. You have emissions from transporting food, but you can also be a net carbon sink and even become carbon negative. This allows our customers to participate in the carbon credit market if they can prove that they are putting carbon dioxide back into the ground.

It helps that plants have figured out their circular economy eons ago: Photosynthesis breaks up carbon dioxide with solar power, so the oxygen is kindly released into the atmosphere for us to breathe and for iron to rust. The carbon and water are then used to make glucose and other hydrocarbons like cellulose for cell walls, dandelion stems, tree trunks. After a plant dies, it decomposes again into mostly carbon dioxide and water.

8.2.3 Heavy Metal and Other Materials

A wide range of materials need our focus to create a circularity that nature is demonstrating for us. Jamieson said:

Metals including aluminum, steel, copper, and definitely silver and gold are more circular than other materials. Glass also is, to some extent. But we have to figure out what we do about so many other materials that currently pile up in landfills or go through the stacks of waste incinerators or cement kilns. In almost every category, we have this demand and opportunity scenario and how. We can explore new circular ways of doing things to drive a transition more quickly and in a more economically viable and interesting way.

Jamieson cited progress in SAP's customer base:

- Clothing retailer H&M has been piloting clothing rental models in their stores in Stockholm, Sweden. While its business has historically been in product sales, the company is experimenting with a new business model.
- Queen of Raw is a marketplace that gives a new lease of life to sustainable materials, including organic cotton, peace silk, faux leather and fur, and quality deadstock fabrics.[9]
- Consumer goods giant Procter & Gamble uses SAP's asset management solutions to improve or reduce the wastage out of diapers. It is looking to make diapers with less plastic and to develop new ways to recycle a classical one-way product.[10]

Matt Reymann, global vice president of the chemicals industry business unit at SAP, agreed with that materials-based approach, and noted that you're more likely to find surprises where you don't expect them. For starters, Reymann said, the chemicals industry gets a bad rap, which it doesn't fully deserve:

> *I've been in the chemicals industry since I started working. Right after school I started with DuPont. The chemicals industry is so important but their image is not very sexy. Some of these companies have been around for more than 100 years and they are reinventing themselves to be around and relevant for the next 100 years. DuPont started by making gunpowder. And now they're working on things like electronics and water solutions—they are continuously evolving.*

He said some chemical companies have started to redeem the industry's image by shifting their focus from products to outcomes:

> *Some of these pioneers who are changing the image of chemicals over the last decade are no longer just pushing products. They are thinking about creating outcomes for their customers and their customers' customers. There's also a sharper focus on what they define as their purpose, why they are—and should be—around. This thinking also changes the industry's image. You hear leaders like Dr. Ilham Kadri, CEO at Solvay, who articulates the importance of chemicals so well. She likes to say 'One eye on the microscope, one eye on the telescope.' We have to be sustainable and we have to be profitable.*

A focus on delivering outcomes instead of just shipping products inspires new business models; we discuss this in depth in our chapter on delivering "everything as a service." In those models, the incentives shift from selling as much product as possible to delivering the outcomes—for example, a coated part—with the minimum input and the best result. Instead of customers experimenting with how to best apply the product, the chemical company can bring their deep understanding of product and process for better quality and less waste, and they should. It's their labs the product came from in the first place.

Reymann said his industry should also be appreciated for the problems it solves, not merely be seen as a culprit:

> *In the past, chemicals have been considered a part of the problem. But they can also be a part of the solution moving forward. I look at the 2021 Nobel Prize in chemistry. It was given to two gentlemen who discovered a new type of catalyst. It took them two decades to be recognized, but we now we have improved ways to make medicine and pharmaceuticals, or even solar cells.*

We like to say "when you point your finger at someone, three fingers are pointing back at you." And this is certainly true when we point fingers at the chemicals industry for the environmental damage associated with them. It's easy to overlook that the chemicals industry makes products that are essential to support our lifestyle. Look around you and (mentally) eliminate everything that comes from the chemicals industry, big and small. Chances are that your old violin and a few antiques will be in a good shape, but only if your home hasn't collapsed over you.

Over decades the chemicals industry has developed a keen sense for the hazardous substances that go into making their sometimes equally hazardous products. The safety standards and procedures applied by the process industries—to which we also count the mill products, mining, oil and gas, and utilities industries—sometimes appear to be extreme from the perspective of a software company, but we rise to their standard when they visit us in our headquarters in Walldorf: We routinely have a short safety moment to remind all meeting participants of the emergency exits and to hold the handrails when they use the stairs.

The process industries have also been the first to adopt the SAP's environment, health, and safety portfolio of solutions in response to increasing awareness and regulation in this space.

8.3 Numerous Players Driving the Sustainability Shift

James Sullivan is vice president of sustainability management and strategy at SAP. He has had a long career in environmental affairs, including stints at the US Environmental Protection Agency and the law firm Skadden, Arps. He added more perspective on the planetary context:

> *Business operates within the boundaries that society sets, and society must operate within planetary boundaries. A chronic violation of these boundaries makes headlines every day, such as climate change or the ocean plastics crisis.*

We have seen disturbing news in the past few months: heatwaves in India, landfills combusting spontaneously—this brings me back to the beginning of the movement that led to founding of the Environmental Protection Agency in 1970. When did people decide to take action? Well, when the Cuyahoga River caught fire in 1952.[11] *All of a sudden, people said, 'This is not a natural occurrence. Maybe we need to begin to do something about this.'*

Sullivan noted that, depending on perspective or circumstance, individuals and organizations are motivated to act by different planetary boundaries:

> *There is more than one planetary boundary: Climate change leads to more extreme weather events with catastrophic consequences. If we continue to dump plastic, there is projected to be more plastic than fish in the ocean by 2050. Deforestation leads to wastelands and species depletion, and contributes to climate change. Crossing planetary boundaries predictably makes people and society react by changing their own behavior and demanding change from business.*

He discussed how stakeholders are stepping up in their "predictable" reactions:

- Regulators have enacted more than 400 extended producer responsibility (EPR) schemes and plastic taxes, often lobbied for by NGOs.
- Investors note a tenfold increase in the number of private market funds for circular economy investing between 2016 and 2020.
- Consumers and citizens prefer sustainable products, so their market share is growing.
- Gen Z and millennials prefer to work for environmentally conscious companies. Sullivan emphasized that employees "want to proudly work for companies that share their values. The best and the brightest look for a meaningful role with meaningful work."

Sullivan cited research on consumers' willingness to spend[12] as indicating mostly indifference to product sustainability in the past. Now, he said, a real shift has occurred, driven especially by the younger cohorts:

> *We now have data that shows that products marketed as sustainable are growing much faster than conventional products in about any category that you look at. What's more, the margins for these products are higher.*
>
> *The data have been robust through COVID, and robust across generations. Generations Y, Z, and younger are pushing change, and they're not even at their full spending power yet.*

8.4 Integrated Reporting and Changing Perspectives

Sullivan described SAP's own journey toward sustainability and circularity:

> *About 10 years ago, SAP started integrated reporting when our CFO Luka Mucic took over as the sustainability champion on our board. Prior to our first integrated report, we did what pretty much everyone else did: We used the same lagging indicators like revenue and profitability. How much money did we make? How efficiently did we make it? What did we do last quarter? What did we do last year?*

After the first report was published, SAP started to include additional metrics:

- Net promoter score: How inclined is a customer to recommend us?
- Innovation score: How innovative are we in the eyes of our customers?
- Employee engagement: How engaged and innovative do our colleagues feel?

These indicators allow us to internally predict how innovative our products will be; how loyal and happy our customers will be; and how engaged, curious, and creative our people will be. Sullivan also found that rich data and deep insights are the key to keeping stakeholders in SAP's integrated reporting happy:

> *They look at sustainability holistically in the context of planetary boundaries, societal reactions, and business realities. They realize that the leading indicators are relevant to assess the potential for future business success. It also provides a forward-looking perspective on what regulators might do next year, what trends will influence customer behavior, and from which angle investors will look at our business.*

The integrated report does more than reassure us that we've done a decent job running SAP in the past year. We know that past performance is not a great indicator for future success. In particular, in turbulent markets where short-term disruptions interfere with longer-term customer and technology trends, the integrated report and the philosophy behind it help us make decisions for the right market position one or two years down the road. For example, over the years, more sustainability-related topics have shown up on our radar, like product carbon footprints, greenhouse gas protocols, commitments to carbon neutrality, and extended producer responsibility (EPR). Clearly, our customers in the manufacturing or consumer industries would be equally—or even more—exposed to these trends, and we shifted product investments accordingly.

But sustainability or circularity trends are not just compliance driven. Businesses are starting their own initiatives. Sullivan shared his experience at the US Environmental Protection Agency when he toured the Ford plant at the start of the Flint, Michigan, water contamination crisis in 2014:[13]

> *One of the solutions was importing millions of plastic water bottles to get people fresh drinking water without lead and legionella contamination. Estimates show that anywhere from 31 to 100 million bottles were generated as waste in the first three weeks alone. This could have led to a massive local waste problem, but local companies stepped in to create new products from this wastestream such as eyeglasses, clothing, and even automobile parts.*

Sullivan provided other examples from the SAP customer base with a focus on reducing emissions in a tiered scope system. He reminded us that scope 1 covers direct emissions from owned or controlled sources, while scope 2 covers indirect emissions from the generation of purchased electricity, steam, heating, and cooling consumed by the reporting company. Scope 3 includes all other indirect emissions that occur in a company's value chain.[14] Sullivan explained:

> *Consumer product companies are getting a handle on their scope 3 emissions and product carbon footprints especially for their upstream agriculture ingredients. Raw materials like palm oil from verified sustainable and non-verified sources often get mixed after the first stage of the supply chain; the origin information will be hidden or lost. Unilever ran a proof of concept for GreenToken by SAP in Indonesia. They sourced 188,000 tons of oil palm fruit from their suppliers, and our solution created tokens that mirror the material flow of the palm oil fruits through the supply chain to track unique attributes and link them to the oil's origin.*[15]

This goes back to our initial discussion about the inquisitive jeans buyer who wanted detailed information about the origin of the material. If Unilever wants to market their palm oil as sustainably grown and responsibly sourced, they better have robust evidence and documentation to support this claim.

Sometimes, for Sullivan, the problem is not so much the product but rather the packaging. For one large beverage company, packaging is the largest single source of scope 3 emissions. For a large consumer products company, it's the second-largest source at around 12%.

If our consumer product customers can determine the greenhouse gas footprints of the agriculture and raw material inputs and packaging, they have quantified a large portion of their scope 3 (upstream) emissions. The emerging standards and related solutions, like GreenToken by SAP, gain customer trust along the entire value chain.

Sullivan pointed to TemperPack, an SAP Business One customer:

> *In modern speak you would characterize them as a company that is disrupting the cold chain with curbside recyclable cold chain packaging—but of course, disrupting is exactly the thing you should not do with a cold chain. Think of*

HelloFresh, the meals that come to everybody's house these days. Think of pharmaceuticals that need to be cooled. A lot of last-mile delivery needs to be temperature controlled, and this is mostly done in boxes made from expanded polystyrene styrofoam. Substituting single-use packaging with recyclable material is a great first step in eliminating unnecessary packaging waste.

Allbirds is an SAP customer running on SAP S/4HANA. This California-based company makes running shoes, apparel, and accessories using planet-friendly natural materials such as merino wool and eucalyptus tree fibers instead of resorting to the synthetic materials commonly used by their competitors.[16]

Our automotive colleagues are involved in the Catena-X network project, which we discuss in Chapter 3 about the Integrated Mobility megatrend. Circularity starts with the design of the cars and the materials used, from the bumper to the battery and from the electric motor to the car's interior. Bringing in all the players along the supply chain creates efficiency, synergies, and new opportunities to increase the circularity of components and materials.

Most people would agree that cars are more complex than food containers, but Sullivan is not fully convinced. He recalled attending a lunch meeting hosted by the National Geographic Society. After he and his colleagues in the sustainability group finished eating their sushi, they tried to be role models by placing the packaging materials in the correct recycling bins:

> *We looked at the packaging that this takeaway sushi came in. The outer box was one type of plastic. The soy sauce came in a little fish-shaped plastic bottle, different material. Probably even the bottle and the little lid were different materials. The packets and sachets for the sauces and for wasabi were other materials again. And finally, the chopsticks. It took a 10-minute discussion to identify the materials and put them in the correct bin, and even the deep experts in our group weren't 100% confident they got it all right.*

We don't have one single solution to solve all circular economy challenges at once, and we agree with Sullivan's friend Mark Buckley, former chief sustainability officer at Staples, who said, "There's no silver bullet to what we're trying to do. There's silver buckshot, and you need to be firing a lot of it at the problem."

We are confident that we could solve the sushi takeout recycling problem with compact handheld near-infrared spectrometers to identify packaging materials. But for decades we've had an even better solution for cafeterias in our Walldorf headquarters: We use silverware and porcelain plates for food, porcelain cups for coffee, glasses for water. The customer cafeteria even features napkins made from linen.

8.5 The Eastman Story

The circular economy is not just about compliance—it's also about opening up opportunities for new products and revenues. Examples come from Eastman, based in Kingsport, Tennessee, which is not only innovating at a fever pitch in its labs, but also deploying promising technologies at scale.

One focus area is the mechanical recycling of plastic waste, retrieved from home recycling bins and sorted at a recycling center. Next, the material is shredded, melted, and re-formed as plastic pellets to make new products. Mechanical recycling works well for items that are marked with resin identification code (RIC) 1 or 2—products like clear, single-use water bottles. The system doesn't work as well with RIC 3 to 7 products like plastic eyeglass frames, which often end up in landfills or incinerators.

In 2023, Eastman will complete construction and begin operation of what will be one of the world's largest plastic-to-plastic molecular recycling facilities at its site in Kingsport, Tennessee. Through methanolysis, this world-scale facility will annually convert 100,000 tons of polyester waste that often ends up in landfills and waterways into durable products, creating an optimized circular economy.

In January 2022, Eastman announced a second molecular recycling plant in Port-Jérôme-sur-Seine in Normandy, France. Eastman is investing $1 billion in the facility, which will recycle approximately 160,000 tons of hard-to-recycle polyester waste per year. In an announcement, the company said, "With the technology's highly efficient polyester yield of 93% and the renewable energy sources available in Normandy, Eastman can transform waste plastic into first-quality polyesters with greenhouse gas emissions up to 80% less than traditional methods."[17]

And most recently, Eastman has announced plans to build a second methanolysis facility in the US that will be able to recycle 110,000 to 160,000 tons of hard-to-recycle polyester waste each year. With these investments, the company is well on its way to meetings its goal of recycling 250 million pounds of plastic waste annually by 2025 and 500 million pounds by 2030.

Aldo Noseda is the chief information officer (CIO) at Eastman. While he oversees IT, he has a much broader mandate: to drive digital innovation at the century-old company, once part of the iconic Eastman Kodak family. Noseda has had a long career in the chemicals industry: He worked at Monsanto, now part of Bayer, for 27 years and can easily explain complex concepts like "mass balance." He used a simple ballpen to ask, "Where does the plastic come from?"

> *It could be from two discarded T-shirts—the old ones with polyester—that account for 20% of the molecules. It could include three bottles of plastics making up 30% of the molecules. The remaining 50% may not be recycled. It could be basic ingredients from petroleum. Now, you don't know exactly in*

the chemical process how the molecules are going to intermix, but you do know, through a mass balance calculation, that the plastic pellets that made this pen are going to have 50% recycled content: 20% come from the T-shirts, 30% come from the plastic bottles, and 50% that is not recycled content.

International Sustainability and Carbon Certification (ISCC) standards are used to certify the mass balance calculations in bulk production. Eastman uses two types of molecular recycling technologies: carbon renewal technology (CRT) and polyester renewal technology (PRT).[18] CRT can process a broad mixture of plastic waste as material sources—in some instances, items as diverse as mixed plastics, textiles, and carpeting. This mix is broken up into small molecules, chemical building blocks that are used to make a broad range of new plastics. Meanwhile, PRT takes a smaller range of polyester plastics, such as soft drink bottles, polyester carpeting, or polyester-based clothing. PRT breaks them into the basic monomers like ethylene. These monomers are then sent through a polymerization process to make the final products, for example, polyethylene.

Noseda explained some advantages of molecular recycling:

> “ *With molecular recycling, you can regenerate the plastics exactly as if you would create them in its raw form. You have no deterioration, no problem with coloration. You can recycle molecules as often as you want. Mechanical recycling is more limited in the number of times that you can recycle.*

Missing capacities are currently the biggest obstacle for large-scale mechanical and molecular plastic recycling, said Noseda:

> “ *We view molecular recycling as a complement to mechanical recycling. Today, there is just not enough recycling capacity in the world, and a big percentage of plastics end up in landfills or in the ocean—just because of the lack of recycling capacity and awareness.*

GreenToken by SAP is a supply chain solution that offers companies a new level of transparency with reliable, blockchain-based information. Noseda is testing the capabilities and said, “We are co-innovating with SAP. SAP has the digital technology and expertise. We have the chemistry and circularity expertise. If we are successful, everyone will benefit from this type of solution.”

Noseda then explained how Eastman is also transforming the world of fashion with its Naia™ Renew cellulosic fiber.[19]

> “ *It’s a pretty cool product. It’s very similar to silk, so it has a luxurious, cool touch which glides smoothly over the skin, even though it’s synthetic. Naia™*

> *Renew is an innovative solution for the fashion industry's sustainability—enabling circularity at scale. The fiber is produced from 60% sustainably sourced wood pulp and 40% certified (mass balance approach) recycled waste plastics. Naia™ Renew is a beautiful product, created from hard-to-recycle waste materials that would otherwise be going to landfills or waste incinerators.*

Molecular recycling and new types of fibers are exciting developments, but what's even more interesting is how Noseda talked about the advanced analytics and ubiquitous computing power redefining research and development and accelerating product pipelines. He described the data science journey at Eastman:

> *First, we did some basic work to get the fundamentals right. We built a data catalog. We established a strong governance. Next, we launched a concept that we called certified dashboards based on a single source of truth. We also ensured that decisions were channeled through a single pipeline, so we wouldn't get distracted by fragmented initiatives. As we were building those dashboards, we found that we needed more advanced models and more sophisticated mathematics.*

Noseda shared a few more insights about how Eastman's attitude toward analytics and data science has evolved as the field matured:

> *We didn't create data science to figure out what to do. We first identified the hard decisions that required complex mathematics and then brought in the data science to help us. A decade ago, analytics was as buzzword. The attitude was, 'Let's play with it and see where it goes.' Now, playing is over, we are very serious about the chemistry and the business, because now we can make real money.*

One of Eastman's biggest successes has been in research and development. Noseda recalls a highly manual process that took nearly a month between the design of a chemical formulation and when it could be tested for effectiveness. Eastman's simulation of the hypothetical lab process shortened the assessment window from months to minutes. For Noseda, this spectacular time savings wasn't the only exciting result:

> *What was even more interesting: the mathematicians were able to reverse engineer formulas. You start with the outcome you'd like to achieve, and the analytics throws out a formulation. The system will tell how probable it is that you'll get the result that you're looking for.*

These digital capabilities are exciting from a scientific perspective. But Eastman can also get its products to market much more quickly. Getting mature data science into labs makes chemists much more effective, a combination of intuition and analytics that translates into real money.

Eastman is also collaborating with Sphera, a leading global provider of environmental, social, and governance (ESG) performance and risk software on their latest iteration of lifecycle assessment (LCA) automation software. This cutting-edge software will enable the scale-up of product carbon footprint and full LCA data across Eastman's entire global portfolio.

For Noseda and Eastman, the circular economy mindset has become integral to the way they do business. Some companies have sustainability departments because they require reporting at the end of the year, but perhaps are not thinking about sustainability as a business. Eastman is thinking about sustainability because it's the company's biggest growth opportunity beyond traditional innovation processes in its specialty chemicals business.

8.6 Circular Economy and Digital Enablers

Jamieson shared his point of view about how SAP fits into the movement toward a circular society. "We don't produce materials. We're not a chemicals company," he emphasized. "We're a technology company that ultimately helps our customers be best-run businesses; we help them do what they are best at."

Jamieson described how data crunched by artificial intelligence (AI) can be used to better manage waste and materials:

> *We take production data of what was sold in which location. Then we use machine learning and geospatial analytics to predict the ultimate route and fate of these materials. We can then go back to our customers' product managers and tell them what will happen to their products downstream. And if they have the budgets in place to make this downstream flow circular, that's where we think they should spend the effort. And if they don't have the budget, our insights help them build the business case to get it.*

This example shows how better insights based on operational data and clever analytics can drive business innovation, discover next practices, and point to promising new business models. SAP Responsible Design and Production is the foundation to design products that fit into the strategic framework of our innovative customers.

But generating cool ideas for circular products isn't enough. Jamieson discussed the guidelines and regulations that companies need to comply with and suggested a framework to address the complete circular thinking across most material flows. He explained his three principles:

- Empower business to eliminate waste by embedding insights into business processes and driving transparency for stakeholders
- Increase the value of materials that can be reused, for example, by marketing and selling materials (that would otherwise be waste) in an open business network
- Shift from linear product consumption to regenerative business models that emphasize reuse to stimulate new usage behaviors

We follow these principles with innovative products for responsible product design and production, and with solutions for responsible sourcing on marketplaces. SAP's industry cloud is an open innovation space with digital enablers for SAP and partners to build intelligent solutions to drive the circular economy.

Jamieson noted that the concept of extended producer responsibility (EPR) has already catalyzed change in the complex landscape of regulations:

> *A good EPR rule will be different in a few years from now from what makes sense today. Technology, infrastructure, and industry practices evolve, new opportunities for circularity emerge, so the policies need to keep up with this change.*

He offered a very interesting point of view on the relationship between business and regulation. He sees significant synergies between sustainability leaders and regulators. Policymakers can create an even playing field with rules for all players. Sustainability champions would have an advantage because of their early or eager adoption, while others will need to catch up with the rules. Thus, rules must set challenging yet achievable goals: Overly ambitious goals will be ignored because no players could possibly comply with them; overly lenient goals don't require any effort and so nothing gets done.

Our role is to make this system work with digital solutions that measure current performance and provide our customers the tools to continuously operate in an ever-evolving regulatory framework.

Jamieson described how insights into materials flowing along value chains benefits the stakeholders:

> *SAP Responsible Design and Production connects policy with business action. We establish a standardized data foundation that enables our customers to integrate sustainability metrics in their design and production processes.*

This standardization creates internal transparency into the sustainability impacts of business decisions, but it also works upstream with the supplier base to influence the sustainability performance of materials and components. But the downstream impact on customer engagements may be even more important because customer preferences and demand for sustainable products are a big lever to influence the speed and direction of market transformations.

8.7 Startup Ecosystems

Beyond SAP's vision and our broad portfolio of sustainability and circularity solutions, we tap into the creativity and innovation of a vibrant ecosystem of startups for whom circular thinking is foundational to their products and services. For example, we've worked with Edinburgh-based data analytics group Topolytics and its WasteMap platform that blends mapping with machine learning. This company is on a mission to make the "world's waste visible, verifiable, and valuable." We have jointly created Scotland's Waste and Resources Map as a showcase that generates a live view of materials that flow into, within, and out of Scotland.

SAP.io is our strategic business unit to incubate startup innovation and to develop and drive new business models for SAP. Our startup program SAP.io Foundries focuses on promising startups on the way to deliver value to SAP customers. These programs run in ten innovation hubs: San Francisco, New York, Berlin, Munich, Paris, Tel Aviv, Singapore, Bangalore, Shanghai, and Tokyo. Each location incubates 15 to 20 startups twice a year, resulting in an impressive 300 to 400 startups that have the opportunity to define and pursue their innovation and growth path.

Several startups in the SAP.io Foundries program cohorts are focused on the circular economy. For instance, Algramo, whose name translates to "by the gram," is based in Santiago, Chile.[20] This company works with consumer packaged goods companies like Unilever to rethink the packaging and refilling of home and personal care products, pet food, and other products usually sold in plastic containers. Algramo is working on a system of vending machines that refill reusable smart containers using radio-frequency identification (RFID) and mobile apps to track consumption, usage patterns, and the container lifecycle. Pet owners concerned about their darlings' healthy lifestyles may also be an attractive target market for a range of pet products and services. A mobile app to track containers may one day become a sales channel for value-added products and services. Reusable containers also make customers come back regularly. Replenishing the vending machines, as performed by Algramo, can save retailers' precious shelf space. What starts as an idea to reduce plastic waste and increase circularity may have impact beyond initial ambitions.

Another example is LimeLoop.[21] In this company's business, retailers ship goods to consumers in reusable, weatherproof, and temperature-controlled containers made from recycled billboard vinyl. Consumers use a prepaid shipping label to return packages to a third-party logistics company that sanitizes and redistributes them.

LimeLoop's containers are designed to be reused over 200 times, whereas the usual cardboard boxes only survive a few cycles. The boxes are equipped with sensors to track and trace shipments, which avoids customer complaints about lost, delayed, or damaged shipments. LimeLoop wants to eliminate many tons of discarded packaging, reduce recycling emissions, and lower delivery costs for retailers.

8.8 From Circumnavigation to Circular Economy

Dame Ellen Patricia MacArthur is a highly decorated sailor. In early 2005, she broke the world record for the fastest solo nonstop circumnavigation. She completed her 27,000-mile journey in 71 days, overcoming an array of bad weather and other technical challenges.

Long-distance sailing requires careful preparation and logistics: You must take everything you'll need with you because you can't just pull over at the next supermarket to pick up the things you've forgotten or run out of. She described how months of isolation on a self-contained boat made her see the parallels to our planet: Earth may be powered by the sun and not by the wind, but we equally need to make do with what we have "on board." Take this perspective, and you can clearly see that our linear lifestyle of "take, make, use, dispose" can't work in the long run.[22]

She created the charitable Ellen MacArthur Foundation in 2010, which has since become a center for research on circular economies. Sullivan called the foundation "the lighthouse for what circularity is." Its iconic "butterfly diagram" illustrates the technical cycle that keeps products and materials in circulation with reuse, repair, remanufacturing, and recycling processes. On the left "wing," you'll see the biological cycle where nutrients from biodegradable materials are returned to the Earth to regenerate naturally.[23]

Much of the foundation's impactful analysis looks at the pathways of different material types. Jamieson explained the vision that the foundation has initially developed for the fashion industry but that can be applied to other material cycles:[24]

- Virgin resources are minimized by increasing the use of existing products and materials.
- By-products are minimized and treated as valuable materials.
- Recycled content is used to protect finite feedstocks and to stimulate recycling.
- Virgin input comes from renewable feedstocks using regenerative production practices.
- Renewable energy is used for manufacturing, distribution, sorting, and recycling of products.

In an interview with Ellen MacArthur Foundation, Solvay CEO Dr. Ilham Kadri discussed a project with Veolia and Renault to extend electric vehicle battery life and increase battery circularity.[25] She explained how difficult and complex it can be to put the simple idea of circularity into practice. Some material recycling from dead batteries has been applied for many years, mostly for plastic, aluminum, and copper used in casings. But the recycling of lithium, cobalt, and nickel has so far not progressed—which is where Solvay is engaging. Developing technologies for the recycling of electric vehicle batteries is a major opportunity as the number of electric vehicles worldwide is expected to grow to almost 120 million by 2030. We agree with Kadri: Today most of the electric vehicles running on lithium batteries that were ever made are still on our roads, so while dealing with dead electric vehicle batteries might only be a small problem today, this problem will grow massively in the next years.

In 2020, our CEO Christian Klein, joined the World Economic Forum's Global Plastic Action Partnership community and committed that SAP would join the effort to remove plastic from the world's oceans.

For Jamieson, removing ocean plastic is just one step in SAP's quest for circular industries:

> “ *This is all about the opportunity we have to work together as a global ecosystem, powered by digital solutions, in order to drive the upstream solutions that actually prevent material waste from ending up in the natural environment.*

SAP has been in the business of enterprise resource planning and management for 50 years. Our customers need SAP solutions that acknowledge and deal with the principle of constrained resources. Only so much money, materials, machines, market, or manpower are available, and we help our customers achieve the best economic and societal outcomes for customers and stakeholders within the enterprise's boundaries and regulatory frameworks. It's no small step to expand this philosophy to a planetary scale, but that's exactly what we need to do as humankind if we want to leave a planet to the next generations that will move them comfortably, safely, and cleanly through space and time.

8.9 Circular Economy with SAP

We close this chapter on circular economy by highlighting some of the solutions in SAP's portfolio that offer key functionality for sustainable transportation and consumer communication.

8.9.1 Sustainability in Transportation

To reach the 1.5-degree global heating target, a 50% reduction of carbon dioxide emissions is required over the next 10 years. With transportation being responsible for more than 20% of global carbon dioxide emissions, the transportation industry is under pressure to reduce its contribution to greenhouse gas pollution. This reduction requires insights into the actual carbon dioxide emissions of the various transportation legs and modes and into the business processes to enable better and more sustainable planning.

- The freight collaboration functionality of SAP Business Network for Logistics enables carriers to report actual emissions for the transportation of goods.
- SAP S/4HANA Supply Chain for transportation management provides optimization algorithms for better resource consolidation (and thus, better resource utilization) and for total travel distance reduction through route optimization.
- SAP Product Footprint Management aggregates emissions from different SAP and non-SAP solutions and provides calculations of greenhouse gas emissions per product.

8.9.2 Consumer Preference for Responsible Products

End consumer demand for fair-trade products is growing, and these customers are also willing to pay more for products if they can be sure that materials have been responsibly sourced, that food has been grown ecologically and sustainably, and that no child labor was involved in making the goods they use.

The following solutions that contribute to food security and sourcing might be part of your SAP-run IT landscape:

- SAP Rural Sourcing Management connects smallholder farmers and suppliers in rural areas with the supply chains of global agribusinesses and consumer products companies. You can manage your sustainability data better through digitally recorded information on producers, their farms, and their communities at every level of the value chain.
- GreenToken by SAP provides transparency for raw materials using blockchain technology. Information about whether a product was sustainably sourced, without child labor, and freely and ethically traded can be easily accessed by customers.
- The SAP Business Network for Logistics material traceability functionality improves transparency, visibility, sustainability, and efficiency by connecting all trading partners in the entire supply chain for tracing of materials on batch level. This application is augmented by blockchain technologies.
- SAP Responsible Design and Production is a cloud solution that enables our customers to fulfill extended producer responsibility obligations, calculate plastic taxes, and honor corporate commitments for responsible material choices.

Chapter 9
Resilient Supply Networks

"The Greatest Show on Earth"

The Ringling Museum in Sarasota, Florida, is an extraordinarily attractive venue. It sits on a waterfront, hosts a mansion fashioned after a Venetian palazzo, and features statues overlooking a majestic courtyard reminiscent of St. Peter's Square in Rome with a lush garden adorned with great banyan trees. The museum's collection includes nearly 30,000 objects of art, including five of the only seven surviving giant canvases of *The Triumph of the Eucharist* by Peter Paul Rubens.

But supply chain professionals don't come here (just) for the art, the gardens, or the waterfront. They seek out the Tibbals Learning Center on campus. Or more specifically, they want to marvel at the 1:16 scale model (spread over 3,800 square feet) of a tented circus during the golden age of the circus in the US during the 1920s and 1930s.

Howard Tibbals spent his entire adult life building this model, which comprises over 42,000 objects, including not only the big top, menagerie, and sideshow tents, but also 59 rail cars, 54 wagons, and over 900 animals. Each tent and each wagon are equipped exactly like in a real circus, from the steam kettle in the cookhouse tent to the 7,000 folding chairs waiting for an audience.

In its heyday, Ringling Bros. and Barnum & Bailey Circus required more than 100 rail cars to ferry about 1,300 workers and performers, plus animals with their cages and equipment. The circus would travel up to 15,000 miles over an 8-month season and visit as many as 150 towns. These days, we build digital twins of physical objects, even circuses. The Tibbals model gives visitors a more visceral experience of the complex logistics circuses had to operate year after year without any of today's technologies:

- The circus might stay at some large US cities like Chicago and New York for a week or two, but most other towns they would visit for only a day. As a cash-only business back then, the tour would start in the bigger cities to build up a cash cushion. Other circuses would go broke in the middle of the season. Of course, no digital payment or cash flow technologies existed during the golden age.
- Loading, unpacking, repacking, and travel all required unbelievable precision. The evening performance ended at 10:30 pm, but the circus had to be in the next town by 4 am so it could start the in-town parade by 11 am. Lots of parallel activities were required: Tearing down and moving stuff to the train yard while the performance

was still ongoing. Cooking for everyone while the Big Tent was being set up, all without any software for project or human resources management.

- Local supply chains needed to be just as precise. A typical daily food order for the circus's people would consist of something like 90 gallons of fresh milk, 1,000 pounds of bread, 300 pounds of beef and pork, 250 dozen eggs and 300 pies of four varieties. For the animals, the order was perhaps more like five tons of hay, 20 tons of straw, 50 bushels of oats, and 600 pounds of bran. Local vendors also delivered ice, water, coal, gas, and kerosene. No digital supplier networks were around to manage all this complexity, and the most sophisticated communications technology was the telegraph.
- No forklifts or cranes existed to do the heavy lifting. Flat cars, block and tackle, and the ingenious use of elephants and horses helped to load and unload animal cages, tents, and cooking equipment. The animals were all-weather, all-terrain specimens—they would work though rain, snow, and mud—and were also part of the entertainment.
- At the industry's peak, the day the circus came to town ranked as a public holiday: Banks and businesses closed, schools got a day pass, and an entire town came out to watch the parade and see exotic animals and acrobats. The shows used large newspaper ads, heralds, courier booklets, and small picture cards, but the bulk of the marketing budget was spent on colorful posters of all sizes and the cost of putting them up. There were no TV ads or social media.
- The planning process began months in advance of the actual show date, with early visitations to the town by contracting agents and the promotional staff who negotiated with local printers and glued thousands of posters around town. Weather routinely played havoc with the plans. No spreadsheets or project planning tools or 10-day weather forecasts were available back then.

The circus didn't survive the onslaught of digital entertainment and gaming or the animal rights movement, but it has been studied extensively. According to an 1895 article in *McClure's*, the US Army sent officers to study Ringling Bros. and Barnum & Bailey Circus for a week while the circus was visiting Fort Leavenworth, Kansas. The officers' report praised the circus's complex logistical operations. Since then, the circus's logistics have been scrutinized in detail in the *Production and Inventory Management Journal*[1] and in many other publications.

Indeed, the near-daily packing and unpacking of so many rail cars, the effort required to manually raise the circus tent, the cooking of meals for the whole contingent, and the town parade were all spectacles that fit the moniker forever associated with the circus: "The Greatest Show on Earth."

9.1 Supply Chain Crises Abound

Shortages of toilet paper, baby formula, and lumber have made headlines. Semiconductor and fertilizer shortages have affected significant swaths of the global economy. The chaos caused by a single ship getting stuck in the Suez Canal for a few days is magnified so that many more commercial vessels were stranded around ports as far as California. Ports closing in China per its zero-COVID policy have sent logistics shockwaves around the globe.

With all this turbulence, even forecasting has gone wildly off track, impacting some of the most established brands in the supply chain community. Richard Howells's regular posts in the SAP BrandVoice series on the *Forbes* website has provided some interesting information nuggets:

- Baby formula is highly regulated and relies on four main suppliers accounting for 90% of domestic US production. According to Howells, the disruption risk "is compounded by the fact that there are no alternative plants to manufacture the product in the US, which has resulted in costly emergency shipments from the closest available source—Ireland."[2]
- In his 2022 post about the semi-conductor shortage, Howells wrote that "In 1990, 37% of computer chips were made in US factories.... By 2020, despite accounting for an estimated 47% of semiconductor sales, that number had declined to just 12%. Europe's share declined to 9% percent from 40% over the same period. Manufacturing is now concentrated in Taiwan, South Korea, and China. So, when several foundries and integrated device manufacturers (IDMs) closed or ran reduced shifts early in the pandemic, it created a noticeable impact on global capacity."[3]
- In another post about shipping obstacles, he wrote that "It's not just the west coast that is affected. There are similar problems up and down the east coast off New York and Georgia.... This is compounded by astronomical rises in shipping costs. It is estimated that the cost to ship a container from southern China to the west coast of the USA is as much as $20,000, up from $3,000 pre-pandemic. That is a rise of over 650%."[4]

We appear to be lurching from one crisis to another. Shouldn't our supply chains have become much more resilient in the century since the circus's heyday, given all the planning and execution technology we now have? Many of these incidents are discrete and have varying points of failure, but with so many businesses blaming poor performance on COVID-19 or Brexit or the war in Ukraine, we are masking the root causes of these problems.

Actually, if you step away from those highly publicized breakdowns, it is remarkable how well many supply chains have performed. Here are a few examples:

- Apple designs its devices in California, maintains a global supply chain to deliver components to Foxconn plants in China, and has managed to deliver over 2 billion devices through multiple channels around the globe. It has done so with constantly evolving bills of materials (BOMs) and networks of suppliers—and still has largely managed to keep its detailed specifications closely guarded secrets.

- Amazon could have easily used the excuse of staff shortages and other COVID-19–related issues that affected every physical and outdoors activity, but instead it turbocharged its e-commerce during the pandemic. Its fulfillment center in Kent, Washington, can process more than 1 million items a day, three times what was possible at the company's state-of-the-art warehouses a decade ago.[5]
- The COVID-19 vaccine supply chain provides other mind-boggling numbers, as a Peterson Institute for International Economics white paper detailed: "By the end of July 2021, roughly 4 billion doses had been administered worldwide. Most required a two-dose regimen—if that trajectory continued, close to 14 billion shots would be needed to inoculate the global population."[6] The BioNTech/Pfizer vaccine required especially complex cold-chain logistics. The company said it had "developed temperature-controlled thermal shippers that utilize dry ice to maintain recommended storage temperature and a GPS tracker for 24-hour location and temperature monitoring."[7]
- In the US, retail returns jumped to 16.6% in 2021 versus 10.6% a year prior, according to a survey by the National Retail Federation and Appriss Retail.[8] These figures add up to more than $761 billion of merchandise that retailers expect will wind up back at stores and warehouses. Most consumers do not realize how complex the reverse logistics can be. A third-party logistics provider, goTRG, processes more than 45 million returned items per year for Amazon, Home Depot, Walmart, and others, using sophisticated software and a fleet of massive warehouses. After integrating the specifics about each retailer's catalogs, goTRG's artificial intelligence (AI) determines the most profitable outcome for every item. Products may be restocked, refurbished, donated, recycled, or used for parts—all within goTRG facilities.[9]
- Mumbai, India, has been known for over a century for the remarkable precision of its *dabbawalas*— couriers who pick up food in lunchboxes (*dabbas*) from the homes of customers and deliver them by train, bicycles, and carts to factories and offices, so that workers can enjoy a hot home-cooked lunch. They then reverse the logistics and deliver the 200,000 *tiffin* boxes back home. Even though the technology is basic, they have a remarkable Six Sigma track record—estimated to miss only 1 delivery in every 6 million.[10] That model has been modernized with mobile apps. In 2020, the Indian online food delivery market was worth around 2.9 billion US dollars.[11]

9.2 Supply Chain Resiliency and Digitalization

These examples show that highly sophisticated digital and analog supply chains optimized for maximum performance at minimal cost are most susceptible to disruption: Redundancy, buffers, safety stock, and backup plans cost money and can hurt a supply chain executive's career—as long as disaster doesn't strike. And if it does, how do you really distinguish between a lack of preparedness and *force majeure*? But perhaps the root cause for the vulnerability of supply chains can also be found in their complexity.

Martin Barkman, senior vice president of solution management for SAP's digital solution management team, provided a historical perspective of how the modern supply chain has evolved over the last few decades and has faced various disruptions:

> *What used to be kind of linear, sequential supply chains where you make it over here and sell it over there, have now become much more intertwined and connected. Supply chains are more like networks than they are vertically integrated, linear, or sequential. That has created a whole new level of dependencies that never existed before.*

Barkman uses an increasingly common phrase to describe what happens when a supply chain can overcome shocks such as the COVID-19 pandemic and war in Ukraine—*resilience*:

> *Everybody kind of jumped on the term, but what it really means is that a supply chain can absorb a shock, continue to function, and then return to an equal or even better state.*

According to Barkman, the awareness of business leaders and the public has finally caught up with the crucial role that supply chain management plays:

> *I think the world has gained an appreciation of the importance of having some amount of shock absorption. It's now top of mind for companies, whereas the supply chain used to be an afterthought, just a necessary component of running businesses—with the notable exception of a few companies that really made the supply chain the absolute core competence of their whole value proposition. For the most part, it was almost treated like a back-office function and a cost element.*
>
> *Now that's not the case anymore, because if you have a good supply chain, you can protect the revenue and profitability stream. It's essential to being able to serve customers the way they want the product. It's essential to being able to serve different types of markets and customer segments around the world.*

Being a sustainable business has moved up the executive agenda, pushed by financial investors, customers, and regulators, but just trying and be sustainable inside your own four walls won't get you far, according to Barkman:

> ***Additionally, you can't even get close to proclaiming that you're a sustainable company if your supply chain is not sustainable. Most of the company's carbon footprint, most of the cost, is typically tied up in the supply chain. So, if you're not thinking of sustainability in the supply chain, then don't even pretend to be thinking of sustainability at all.***

In terms of importance, the supply chain has advanced from the back office to the management team to the board of directors. In many countries, critical supply chains have become a matter of strategic national importance. Certain industries, and the supply chains behind them, are covered in presidential briefings. Supply chains are central to the economic strategy of countries like China and are covered in the mainstream news.

For many organizations, making supply chains more resilient will require some new thinking and new technology. Barkman emphasized that "if you're serious about moving away from previous 'lean' or 'just-in-time' thinking to becoming resilient, you have to think about digitalization." But that goes well beyond merely taking an analog process and digitizing it:

> ***When we say digitalization, we're talking about leveraging the vast amount of information and data to make fundamentally different, better, and more effective decisions, and weaving that into the fiber of how the organization operates.***

He recommended employing a broad definition of supply chains—encompassing everything from design to operations—and a focus on four dimensions: agility, productivity, connectivity, and sustainability.

9.2.1 Agility

Agility includes anticipating and sensing market dynamics and responding to demand or supply disruptions without negative consequences for profitability or service levels. This approach requires a platform that consolidates information around the globe about demand and supply across all markets, uses predictive analytics and simulations to anticipate demand and supply patterns and disruptions, and proposes fulfillment options and alternatives. Such a digital real-time platform is a critical asset for state-of-the-art supply chain operations.

Barkman provided details about two consumer product companies: Lamb Weston and Kellogg's. Lamb Weston supplies frozen potato products and dehydrated potato flakes to food service and quick service providers, industry, and retail customers in more than 100 countries. This company uses SAP Integrated Business Planning for inventory planning and sales and operations planning and reports multiple business benefits, such as the following:

- Automatic calculation and advanced planning models reduced safety stock by 18%.
- Precise, automated calculation freed up planners to focus on analysis.
- Scenario planning and simulation produced a better understanding of the factors that influence the necessary levels of safety stock.
- Integrated planning improved decision-making on order volumes, supported by insights into the financial impacts, results in better coordinated sales forecasts, inventory, and operations planning.

Kellogg’s produces breakfast cereals and convenience foods, including crackers and pastries, and markets its products through several well-known brands. This company standardized its project and product management processes with SAP S/4HANA and SAP Product Lifecycle Management, which allowed the company to move to a single European project, process, and product management system for various brands with alignment across product categories and clusters. This approach also enabled Kellogg's to implement a single initiative management process that provides reliability to the business through better visibility into current and proposed initiatives.

9.2.2 Productivity

In its simplest form, *productivity* is defined as the amount of output you receive from inputs, usually expressed as a ratio. If that were the full story, however, many dissertations and scholarly books would never have been written. In reality, product and production managers must deal with constraints and requirements that make striving for productivity difficult: Product quality must still be high and consistent, lot sizes must be small (why not a lot size of one?), and change requests should be allowed until well after production start. Digital production processes using Industry 4.0 technologies promise to overcome the trade-off between small unit costs resulting from large batches of uniform products versus high unit costs resulting from individualized products. The production processes of premium automotive brands are finely tuned for efficient production of mass-customized vehicles. The number of variants that can be assembled on a single line is astronomical.

Productivity also suffers from unplanned or unnecessary downtimes of production machinery. Intelligent equipment feeds sensor data into even more intelligent analytical systems to monitor machine health and performance. Predictive maintenance organizes service calls and spare parts before a defect occurs or quality suffers.

Barkman cited the automation at Ansaldo Energia S.p.A., a power plant engineering and construction company based in Genoa, Italy. This company designs, builds, and delivers power plants with gas or steam turbines that feature thousands of sensors. The power plants at customer sites are continuously monitored to detect anomalies and correct them before productivity issues surface. Here, maintenance is an integral element to ensure plant safety and productivity.

9.2.3 Connectivity

Real-time information flowing along supply chains through *connectivity* maximizes the lead time to respond to supply disruptions upstream and to demand spikes or drops downstream. Manual processes, the lack of data format and exchange standards, data security concerns, and the fear of disclosing proprietary information all slow down information flow.

Thus, a digitally connected, real-time supply chain begins with business relationships, continues with technologies that address these concerns, and ends with systems that take effective action based on the incoming data, including talking digitally to suppliers about anticipating an impending supply risk or talking with customers to collaborate on recovery scenarios that may arise from supply disruptions.

Information without action is useless, and action without information spells disaster. If you have children, you may find either a severe shortage or a severe oversupply of schoolteachers when your kids enter primary school. This mismatch is surprising since it takes the same number of years for a new teacher to be trained as it does for an infant to reach elementary school: Take a headcount of newborn babies and first-year college students on the track to become teachers, account for retirement and attrition, and you'll know whether a problem will arise 5 years or so down the road.

Connectivity and data alone aren't sufficient. But if you add insight, collaboration, and effective action, you have the key ingredients to make your supply networks more robust via digital connectivity and intelligent systems.

9.2.4 Sustainability

Supply chain disruptions are caused both by mismatched demand and supply and by incidents along the supply chain, like unplanned production or logistics issues caused by faulty machinery, fires, quality problems, strikes, bankruptcy, power outages, wars, embargos, traffic jams and accidents, volcanos, tsunamis, or earthquakes. We're sure you can extend this list with additional examples.

But maybe you haven't thought about supply chains disrupted by sustainability-related incidents. Sustainability-related regulation can cut off the supply of goods and materials that are not produced responsibly and according to safety and fair labor standards. Suppliers may decide to discontinue products if they can't profitably fulfill extended regulations.

Thus, making supply chains more sustainable by eliminating waste, emissions, and inequality also contributes to making supply chains more resilient against related regulations. Businesses that can demonstrate a sustainable supply network will attract financial investors and customers.

SAP has committed to the goal of eliminating emissions, reducing waste, and pushing for equality—not only in our own operations but also all along our supply chains.

Barkman discussed Smart Press Shop, a joint venture between Porsche and Schuler that applies Industry 4.0 principles. The company makes car body parts from steel and aluminum in Halle (Saale), Germany, for highly specialized automotive manufacturers. The goal was a fully digitized production line to achieve productivity targets while offering production in small batches—a key differentiator for original equipment manufacturers (OEMs) with customer-centric, individualized production. Smart Press Shop also pursues ambitious sustainability goals. They are working with the SAP application management partner Syntax to realize a vision of 100% traceability in its production, from raw materials to finished parts, allowing the company to provide footprint and circularity information about its products to customers. The company modernized and automated to operate in a completely paperless environment with systems integrated end to end. Smart Press Shop runs and integrates everything from its production line equipment to its cloud ERP systems, all along the production chain.

9.3 The Maturing of Industry 4.0

Hannover Messe, the world's leading industrial trade fair, celebrated its 75th anniversary in May 2022. The term Industry 4.0 was first used at the event in 2011. Industry 4.0 is no longer a mere buzzword—automated, digital factories are already up and running and keep maturing. And the impact of automated shop floors ripples along supply networks all the way to participants far removed from manufacturing industries. Germany in particular has been pioneering Industry 4.0 on its manufacturing shop floors in recent years. From the shop floor, intelligent, sensor-studded machines connect to suppliers and customers along logistics chains and maintenance functions. Industry 4.0 is now delivering on the promise that more data can result in smarter decisions.

At Hannover Messe, SAP's booth is an event fixture, and we use the trade fair to show off our solutions and many of our industrial partners' robotics and other technologies. This year, SAP presented how digital technologies can be applied in the design-to-operate cycle for discrete manufacturing[12] and for process manufacturing.[13] Hannover Messe is an event for a few days, but customers visit us year-round, so we decided to place permanent Industry 4.0 showcases at our corporate headquarters in Walldorf and in major SAP locations around the globe. The showcases are great conversation starters, with digital technology working hand in hand with physical machines to make innovation tangible and spawn creative discussions.

Rakesh Gandhi is an "innovation evangelist" at SAP who leads tours of the company's showcase location in Newtown Square, Pennsylvania. He routinely interacts with customers and reported that they're looking to incorporate more technology to come away with better data:

> *Customers are asking us, 'How can I run and monitor my operations from a control tower? I shouldn't need to walk around on the shop floor with safety gear and a helmet. I should get push notifications triggered by events and exceptions in production and the supply chain. For example, for production planners, if a supplier can't deliver or a shipment is delayed, what are my options for corrective actions?'*
>
> *We are connecting the enterprise business (IT) processes to the operational technology layer. Industry 4.0 started with machine-to-machine communication. Now, we are communicating enterprise IT processes directly with the robots, using the right exception handling procedure for specific situations. For example, if an electronic board fails a test on the assembly line, I need to push it out to rework and if this impacts a customer commitment, the system triggers an alert.*

Several discrete and process manufacturing tracks exist in the tour. The use cases on the discrete manufacturing track include the following:

- Digital logistics: Outbound delivery with transportation management
- Intelligent workers: Gesture-based interfaces
- Digital twins: As-manufactured twins of the final manufactured equipment
- Quality: Automated testing and quality management
- Intralogistics and the automation of production material flow: Automated warehouses, automated conveyors, robotics integration, dynamic routing, and modular manufacturing
- Intelligent factory: Final assembly of variants, 3D work instructions, digital product history, and labor certifications
- Manufacturing insights: Plant and overall equipment effectiveness (OEE) insights, global insights, and dashboard designers

In the demo, SAP's Matt Ruff plays the role of an assembly station operator who is responsible for the production of a component. He must analyze process data and handle quality issues detected by the equipment. During his demo, he shows how SAP Digital Manufacturing Cloud seamlessly integrates with robots, handles materials movements, automates material flows, controls smart tools that assist workers with assembly operations, and uses smart vision systems in material handling and quality control. Cameras watch the assembly process and perform automatic visual inspections based on AI and machine learning models. His demo script guides customers through the physical and digital experience:

> *You saw a visual inspection of the circuit board. We took a picture to check the wire connections. The inspection passed. We'll see the result of this in the dashboard here. The robot went ahead and picked a casing of the specified color for the final product that we're making here.*

In the next stage of the end-to-end demo, Ruff's colleague Sujit Hemachandran showed off his warehouse robots, called autonomous mobile robots (AMRs):

> *They agree on who best takes over which task. We only have two of those and we have given them nicknames, but you can scale this to a fleet of 100 or 200 robots. If you have seen an Amazon warehouse, there's a lot of them zipping around, communicating and working with each other.*

On the process manufacturing segment of the SAP showcase tour, Ben Hughes described its digital twin capabilities:

> *When I was standing in front of the mixer, you saw it's a big piece of equipment with lots of valves and pumps and sensors. We have that entire equipment structure with its bill of material modeled in our solution that mirrors its physical structure here in the plant. We also have a digital twin of my plant and the mixer is also part of that. The same structure is also what exists in my system of record—my SAP S/4HANA system, my plant maintenance system. That's where I'm configuring work orders that then go directly to the plant and the mixer. We're fully integrated with that.*

The tour shows off the industrial partnerships SAP has formed that have matured over the last decade. The demo in Newtown Square featured xPlanar tiles from Beckhoff, which use magnetic levitation (maglev) technology to move products around like an automated guided vehicle (AGV) would. Daymon Thompson and Jeff Johnson at Beckhoff described how the push for flexible manufacturing and personalized products have popularized the concept of a "lot size of one" in the logistics space. This concept has led to an explosion in robot designs. According to Thompson, Beckhoff's customer base includes numerous machine builders, so it has decades of experience providing advanced machine controls, servomotors, mechatronics, IoT solutions, and more. The machine builders work with end users to figure out how to deploy next-generation machine designs that are more flexible and provide more throughput.

Thompson described what is involved in producing lot sizes of one:

> *The idea is that anyone should be able to order customized products online. Modern e-commerce systems automatically push orders from the website into a manufacturer's ERP system. Then it gets queued automatically for manufacturing. Now, the machine has to be flexible enough to make one-off items without losing efficiency in lot size of one production. Downstream, you need the automation and digitization in the shipping products as well. Automated warehouses, automated storage and retrieval of orders (AS/RS), and automated palletizing and freight loading are just some examples.*

Johnson also suggested to watch out for the digitalized "new normal" in shipping techniques the next time we go grocery shopping:

> *Say you're in Target or Walmart. You'll see that the folks doing shelf replenishment push carts that seem to contain a whole bunch of random products. What you don't see: they didn't go into the back room and go, 'Okay, I need to pick two of these, three of these.' The cart was actually shipped to them that way from the main warehouse. Automation is behind exciting advancements in the intralogistics industry that drive increased efficiency in modern retail and e-commerce processes.*

Johnson shared several other examples: the food preparation industry, pharmaceuticals micro-fulfillment, and modular construction. In the first two industries, customization powered by automation has changed the playing field. Meals can quickly be individually prepared with individually specified ingredients, such as all organic or for special diets, driven by the online ordering process. In pharmaceuticals micro-fulfillment, staging is applied to customize individual orders.

Thompson's example of modular construction concerned prefabricated kitchens. Nobilia, a customer of SAP and Beckhoff, makes 2,500 kitchens a day, some with as many as 1,200 individual parts. They are configured to order and composed and assembled from prefabricated kitchen cabinets and cupboards:

> *When a piece of a custom cabinet comes in, they can identify which part of the order it belongs to. There are hundreds of coordinated conveyors, and they move these pieces around the factory to the different workstations to fulfill each unique order. There are two or three machines that drill the holes for the hinges in the cupboards in the exact configuration specified by the customer. The system also knows which machines are currently available and what their backlog is. It figures out the optimal routing for each part that comes along to*

maximize throughput throughout the factory and to make sure that all orders are completed just in time for delivery.

The only way that Nobilia can handle that kind of production volume is through intelligent automation that avoids backlogs piling up or manually moving parts between the different workstations. This sophisticated manufacturing planning and execution requires robust backend systems with real-time shop floor integration that creates visibility and intelligent responses to shop floor disruptions.

9.4 The Evonik Story

Evonik, headquartered in Essen, Germany, is a global leader in specialty chemicals. Its tagline is "leading beyond chemistry," and its vision is to make "towels fluffier, mattresses bouncier, animal feed healthier, medications more effective, and tires more fuel efficient."

Thomas Meinel, senior vice president and the head of indirect procurement at Evonik, talked about the evolution of business processes for Evonik's business, with over 35,000 suppliers in some 100 countries. By their very nature, procurement processes—like order and fulfillment processes—don't run under your full control or solely within the boundaries of your enterprise. They involve many external entities including suppliers, customers, and logistics and financial services providers. Digitizing this category of business processes and making them more efficient and intelligent can be approached in two apparently simple phases: First, get the end-to-end processes within your enterprise under control; then you can digitally connect them to the outside world. In theory, this approach sounds simple and obvious, but Meinel points out—in almost philosophical terms—that the reality is a lot less clear:

> *I used to run the business process organization for Evonik and we would ask: Is it source to pay? Is it order to cash? Where does it start? Where does it end? The reality is the end of one process is also a beginning for another. It's hard to talk about end-to-end processes because you do not really have an end, much less two.*

Meinel also discussed the move from a transactional mindset to a network mindset, shifting from conceiving of your processes within the four walls of your enterprise to considering its full integration with suppliers and partners. He mentioned the changing significance of ERP systems as well:

> *Before you think about networks, you have to do your homework. Our homework took several years and involved moving from a heterogeneous system landscape to SAP S/4HANA. We have been on it for a couple of years, and it*

supports more than 95% of our global revenue. Half of our employees have regular access to this ERP system, and I call it our digital core. However, an ERP system was the center of the world in the past. Today it still is a kernel, but it's not everything. The world is becoming bigger. We have to integrate with others much more than we did in the past.

Such a degree of inter-enterprise integration works best if you run a robust platform. Meinel explained the role of Evonik's trading platform and how it keeps evolving:

> *You realistically cannot afford to build one-to-one connections with each supplier. SAP Ariba has allowed us to replace the one-to-one relationships with a trading platform. We connect much faster with suppliers. We also have more transparency.*
>
> *The next phase is even more important. You should be able to add and replace suppliers more easily. It's not just about better exchange of data and making processes more efficient, but you're also more adaptive in changing and adapting your network to new requirements.*

SAP Business Network has enabled Evonik to keep its supplier base aligned to its business requirements, but digital transformation must also be created for the broad range of products it needs to run its business. For Meinel, best practices for managing a supplier base are evolving, and sourcing and procurement have become more strategic, now prompting four major ways of interacting with suppliers:

> *We used to buy physical products like pumps. In the new world, we often buy digital products. In the past, I knew my supplier base and it didn't change much. The world was pretty stable. Not anymore: we have many more crises, technology shifts, and so on. So, we need more than a Google search to find and qualify new suppliers, we need intelligent scouting capabilities. We also have a lot more data and need more automation.*

Scouting capabilities became more relevant during the COVID-19 pandemic, especially around indirect supplies. Meinel said he experimented with supplier discovery tools, and Evonik is cultivating a more diverse supplier base with a focus on minority and women ownership, in particular in the United States. Procurement has come a long way and is now a strategic component of the enterprise.

> *Being transparent and compliant isn't good enough any longer; we also need to be fast and efficient. Finally, it should allow me to create next business practices and take advantage of new procurement options, like paying per use*

for a piece of equipment instead of buying it and having it sit on our balance sheet. And then we have to look how we buy combinations of products and services.

Now, most enterprises are connected to their suppliers on one side of the enterprise and to customers on the other. Thus, having procurement, product management, and sales work together can create exciting ideas and opportunities for networked business relationships and processes.

Meinel discussed the challenges he faced in building the digital network around Evonik. Some data- or process-related items don't look too difficult until you try to address them: "How do we ensure one ID for Evonik and all its global subsidiaries?" The change management challenge is even harder to master, according to Meinel:

> *My employees in sourcing may say 'I already know my job. I know my suppliers. How will this help?' And unless there is internal commitment and tailwind from the business, it is also difficult to convince suppliers to comply with new procurement processes. And it definitely takes work. Suppliers have to maintain their data, sometimes they have to requalify. But the platform also delivers efficiency and transparency to suppliers, so there are benefits to their business as well.*

The SAP Ariba platform covers indirect suppliers. But how about direct suppliers and industry networks like Catena-X? Meinel discussed Evonik's caution when it comes to sharing information with others over the network. It never exchanges pricing or volume data but stays on the "information" level, which doesn't raise concerns about disclosing financial, competitive, compliance, or intellectual property information.

But even without exposing price and volume data, Meinel explained a successful example of information sharing with others via Together for Sustainability (TfS) and its partner EcoVadis. TfS is a sector initiative created by chemical companies with the goal of assessing, auditing, and improving the sustainability practices within their global supply chains. Since its founding in 2007, EcoVadis has grown to become the world's largest provider of business sustainability ratings, creating a global network of more than 90,000 companies with a positive rating. Meinel described the relationship:

> *We share sustainability data with EcoVadis as a neutral third company. Evonik, BASF, and all the other major chemical companies do this, and a very high percentage of our direct and indirect suppliers participate as well. We don't share business-critical information like the specific chemistry we buy. We know a supplier is part of the TfS community, but not what they are supplying to our competitors. That is anonymous.*

Meinel highlighted additional potential for more standardized specifications in a chemicals industry that is continuously innovating:

We don't have many raw materials. But we make many intermediates and even more end products. And our units of measure are diverse; we ship truckloads, drums, and thimbles. We start at the beginning with the few major raw materials that we buy in huge quantities, and then go step by step to more—and more specialized—products and smaller quantities. And we need to speak one language with customers and suppliers. So standardized product descriptions are very important.

9.5 New Supply Networks

As senior director of corporate business development, Hari Ashvini leads SAP's engagement with emerging partners. Recently, he has worked with a variety of startups in the digital supply chain space. In the past, when goods left the warehouse, you had limited options: In the US, the main choices were FedEx and UPS. Today, however, numerous logistics services offer low-cost and high-flexibility and often a range of value-adding logistics and supply chain services, including warehouse capacity as a service, commissioning and packaging by the picked item, and rental schemes for scalable warehouse robotics.

Ashvini ran through a gamut of supply chain ecosystem topics: on-demand warehousing, on-demand logistics, warehouse robots, and supply chain visibility. The common theme is that none of these services can be imagined without the extensive use of digital technologies and real-time integration between SAP systems, warehousing infrastructure, and robotics. Beyond more resilience and efficiency for individual components of the supply chain, networking between shippers, customers, and logistics providers with value-added services can make a big difference.

9.5.1 On-Demand Warehousing

Micro-fulfillment moves goods to smaller locations, closer to customers. Some 60% to 70% of the warehouses in the US are owner operated, and companies like StorD and Flexe have capitalized on the micro-fulfillment trend, signing up thousands of companies to run as a single gigantic, distributed warehouse. So, these companies require accurate views of how much space is available, right down to the bin and aisle. The result is that a warehouse will always exist close to the customer, and the shared capacity can absorb peaks, which means an individual warehouse can be operated closer to 100% capacity.

9.5.2 On-Demand Logistics

Similar to the warehouse space, 70% to 80% of freight trucks are owned by operators, noted Ashvini. A truck that sits in the owner's parking lot doesn't generate revenue, and truckers on the road have little visibility into the network about where to get the next load. Expanding beyond passenger service and food delivery, Uber has done a great job connecting shippers directly with carriers. SAP customers who use our transportation management systems in combination with SAP's logistics network are connected to Uber freight to request or offer transportation services and track fulfillment.

The Uber app shows the driver the available loads with their properties like the need for temperature-controlled transportation or hazardous materials classifications—and of course the pickup and delivery location and time. The truckers decide whether they want to bid on the load. Uber also offers truckers its Freight Plus program that provides carriers with discounts on tires, mobile phone plans, and more. When SAP started working with Uber, it was orchestrating some 350 loads a day. Now, business has scaled to 20,000 or 30,000 daily loads in the US alone. SAP solutions also support other logistics service providers like InstaFreight, Loadsmart, or Freightos. The digital connection between shippers, carriers, and service providers drives up asset utilization, increases overall system efficiency, and enables shippers to make—and keep—better service offerings to their customers.

9.5.3 Warehouse Robotics

Robots don't sleep, they don't need breaks, and they can do all the heavy lifting. The SAP warehouse robotics console acts as a gateway to connect multiple warehouses and robots. We have been working with Locus, which offers warehouse-robots-as-a-service teams. Work orders go from SAP system to workers and their robots who work together as "cobots." Locus is collaborating with StorD—so Locus customers don't need their own warehouses but instead purchase high-performance, scalable warehousing services. SAP is also working with robotics companies like Fetch and MiR, in Germany.

9.5.4 Supply Chain Visibility

Ashvini points to SAP's marquee partner project44, which provides multi-modal supply chain visibility around most of the globe to track trucks, trains, ships, and shipments from container loads to the item level—with an accuracy of a few meters. Over the years, its dataset of historical data and the range of real-time data feeds has become extremely rich. We also work with other visibility vendors like Descartes and FourKites. For many years, supply networks have been something of a black hole: You dropped goods in and hoped that shipments would emerge at the right place, at the right time, and in the right condition. Tracking in real time illuminates the network and is a game changer for resilient supply networks that can recover quickly from disruption.

9.6 When the Chips Are Down

When customer expectations and customer commitments for on-time delivery are not met, providers everywhere reflexively reach for the most convenient excuse, one we've all heard before: supply chain disruptions beyond our control. It's true, some disruptions can be considered *force majeure* with no reasonable backup plan in place.

But some disruptions of critical supply chains are self-inflicted, often the result of overconfidence in market forces and in one's own ability to quickly repair supply chains that are damaged by demand/supply mismatches and changing markets or technologies. Jeff Howell, head of SAP's high-tech industry business unit, laconically commented on the recent chip crisis: "This is not the world's first chip shortage and it's not going to be the last. And we are quickly entering into a potential for excess inventories due to a slowing economy, as after two years of a semiconductor shortage, the industry is now experiencing excess inventories." He wrote a white paper about maintaining business continuity amidst recent semiconductor shortages,[14] expanding on the situation in the automotive sector:

> *Demand is out of phase with supply, which has an impact beyond the high-tech industry. The semiconductor industry accounts for 0.3% of US GDP, yet semiconductors are required to produce 12% of the US national output. It's even more pronounced in Japan: 0.4% and 30%, respectively. This is hardly a surprise in a country where every toaster comes with more computing power than Apollo 11 had to their disposal.*
>
> *The auto industry is one of the fastest growing markets for semiconductors, but this sector is dwarfed by the 5G equipment and all the other high-tech products—especially as demand for work-from-home equipment spiked in 2020. In expectation of plummeting demand, automobile companies cut their demand forecasts at the beginning of the pandemic by up to 20% and cancelled orders. Four months later, they came back and wanted to ramp up production again. Unfortunately (for them), the chip production capacity had been sold to other customers, so they had to go to the end of the line and wait for their turn.*

Howell noted that this circumstance brings up a classic risk-benefit assessment of how to use limited resources, and he expects that the market and the supply chain may fix itself sooner than expected. He explained that, in the supply shortage, car manufacturers tried to overcome the conflict between maximizing profit and keeping their biggest customers happy. While Daimler could only get so many chips, those chips could be used for both C-class and S-class vehicles. The company decided that, if a limited number of cars could be made, it was better to build the fully-specced cars with all the expensive, profitable extras and when in doubt, make more luxury S-class cars than the (comparatively) humble C-class vehicles.

Daimler sacrificed revenue but protected its profit. And the sudden shortage of its budget models made evening news and drew even more attention to the self-inflicted chip shortage. Here's what Howell expected:

> *Supply's going to catch up. We have eight companies who have pledged a half a trillion dollars in capital spending to expand their capacity. If the industry keeps building at that pace, we may end up at a point—and it's hard for people to imagine this now—where we have a glut of capacity.*

Howell explained the long and complex semiconductor value chain in more detail. Chip manufacturers are working along three time horizons in parallel to overcome the current chip shortage:

- In the immediate future (up to 6 months in the future), chip manufacturers are focused on yield management, smart testing, multi-tier visibility and exception management, and engaging with the suppliers and customers already in their networks.
- In the near-term future (6 to 12 months away), they are focused on detecting plant maintenance needs, objective-driven customer allocations, and extending to *n*-tier suppliers and foundries in their network.
- In the long-term future (12 to 18 months away), they are focused on reshoring planning to increase domestic capacity, match set optimization, optimizing the *demand forecast-to-order promise* process, and enabling secure network-wide data exchange.

Howell said that focusing on operational efficiency is no longer sufficient and stipulated what he calls a *trifecta of elements*—collaboration, visibility, and insight—across a network of trading partners. Networks like Catena-X for the automotive industry will help with the collaboration. In his white paper, Howell cites ZF, which provides drivetrain, chassis, and safety systems to the automotive industry, and Wolfspeed, which makes wide-bandgap semiconductors for automotive manufacturing as a major market.

Regarding analytics and data-driven insights, there is no shortage of data to analyze; the bottleneck is people. As Howell explained:

> *You just don't have enough (human) capacity to look at all this data. The production of a single wafer generates 15 gigabytes of data. Multiply this with 1,000 wafers per week and 52 weeks in a year, and you're soon dealing in petabytes. So data is not the issue but the effort of crunching the data to squeeze out insights. And we estimate that the industry is looking at maybe 20% of the data that's generated.*
>
> *This is an optimistic estimate. The point is, the data is looked at only when there's a problem. The real questions should be, 'What secrets are lurking in*

the data that we're not looking at?' and 'What decisions would we make if this information was presented to us proactively?'

Howell also talked about the work SAP is doing with its partner PDF Solutions to improve yield management. In this context, PDF stands for Probability Density Function, not for Adobe's portable document format, and hints at the kind of analytical services this company provides. It looks at parametric data and tries to balance tolerance with desired yields: Overly restrictive tolerance thresholds mean lower yields, while overly relaxed tolerances result in more defects—a classic type I versus type II error tradeoff. For readers who like statistics, manufacturing the latest generation of computer chips takes 1,000 production steps and over 3 months. What is the required quality level per manufacturing step to get a 70% yield at the end?

9.7 From Farm to Fork

But now, let's explore a completely different supply chain that doesn't feed the world's demand for digital gadgets but rather, its population.

Anja Strothkämper, vice president of SAP's agribusiness and commodity management solutions, talked about the supply chain from "farm to fork" in the context of current geopolitical conditions:

> *It's a combination of different crises. We see there is the Ukraine war impact. There are still lingering ramifications from the COVID pandemic. Shipping cost has gone up. Nitrogen-based fertilizer is also linked to the price of natural gas, which has shot up and driven production down. A lot of fertilizer comes from Russia and has been sanctioned. There is serious concern that there will not be enough fertilizer for the next harvest year. All of us have become familiar with the term 'bread basket' as a result of the Ukraine crisis. What if there is grain shortage from Russia and fewer sunflowers from Ukraine?*

Most companies don't have visibility two tiers deep into their supply chain. Better visibility is the first step for dealing with supply network disruptions. To try to understand your exposure to a port lockdown in China, for example, you would need to understand who the supplier of your supplier is. As Strothkämper noted, the traceability of products along the supply chain has become even more important:

> *You want to understand the transport and logistics chain to know what happened to a particular lot of grain. But you also need production data from the field: What crop protection or fertilizer was applied? In what conditions does*

> *the wheat grow? What do climate and weather patterns look like? Those are factors that determine crop yield and quality.*

Strothkämper noted that a new breed of consumers has been having a major impact:

> *The consumer behavior is changing. Consumers consistently say they want to know what fertilizer and crop protection was applied to their food (if any). This is driving change in the way that retailers and consumer good companies order products. There's a new generation of consumer, a new sense of what is important to me, and it is driving innovation all along the food chain.*

These concerns illustrate two conflicting challenges in the food supply chain: the sheer quantity required to feed the world and the need for sustainable farming practices to produce food that qualifies as organic. Digital technologies can overcome this challenge. Satellites can provide near-time imagery of farms and fields to determine where fertilizer, irrigation, and crop protection are required—not too little, not too much. Satellite and sensor data digitally and remotely control farming equipment for the best quality and the highest productivity with the smallest possible environmental impact.

9.8 From Reactive to Resilient Supply Chains

Barkman summarized many changes he has seen in companies that are now considered supply chain best practice leaders. These companies realize the value of shared design platforms and networked supply chains:

> *Collaborative design gives you the ability to quickly and digitally swap one raw material for another, try different components, rapidly evaluate variants, and hand over the final design to the sourcing and production stages.*

A networked relationship with many suppliers and quick, digital onboarding of new suppliers enables a seamless way of doing business with a manufacturer's supply base. Automated manufacturing accumulates information to drive for a new level of productivity, he said:

> *This is not just about economies of scale but also about doing things differently. And maybe you can make 'configure to order' products as cost-effectively as batches of a standard product because you're using the information to go to a lot size of one. The result: your plant is operating in a much more digitalized way along more intelligent paradigms.*

Barkman compared reactive with resilient organizations as follows:

Reactive organizations:		Resilient organizations:
Insufficient design insights	→	Collaborative design
Restricted suppliers	→	Balanced suppliers
Increased manufacturing overhead	→	Manufacturing automation
Excessive safety stock	→	Optimized inventory
As-required maintenance	→	Predictive maintenance
Ad-hoc service providers	→	Verified service networks

This checklist enables a quick self-assessment and provides pointers to address the areas where best practices and digital "next" practices can make a transformational impact on the business. Algorithms and AI will play a major role to increase the responsiveness and resilience of complex supply chains.

Barkman said that evolving algorithms are allowing companies to create resilience without additional cost. Inventory management and maintenance can substantially benefit from technology:

> *If you can do maintenance in a more predictive way with IoT-enabled sensors, you can anticipate when maintenance is needed as opposed doing it reactively, resulting in downtime, or just preventively with sometimes unnecessary work and service parts expenses. You could even then use third parties to more easily come in and do the maintenance on that supply chain. For businesses that sell equipment: perhaps they can offer their equipment as-a-service, so that activates the whole servitization theme. They can say to their customers, 'Look, we'll make sure that this equipment is performing, and we'll do the maintenance proactively to keep it up and running.'*

That's how this digitization comes into play. Instead of doing things with just more money, time, and complexity, industries need to start building new platforms and finding new ways to do more with less. As Ashvini explained, "There are endless opportunities to rethink and rewire supply chains for efficiency, transparency, and resilience."

9.9 Resilient Supply Networks with SAP

In a "business as usual" environment, you probably don't even wonder what needs to happen behind the scenes to keep supermarket shelves stocked, gas stations supplied, assembly lines busy, and online purchases delivered to your doorstep. Small disruptions may delay a package by a day or two, or result in your favorite mustard being out of stock on the shelf. But big and sudden supply or demand changes show us how delicately balanced our global supply chains must be to operate reliably. One big push can unleash chaos. So for our peace of mind, we need to be prepared to deal with the shock, and SAP solutions will do their part.

9.9.1 Supply Chain Disruptions

The COVID-19 pandemic caused many disruptions along the supply chain. Agricultural commodities couldn't be loaded and shipped but weren't delivered on time due to traffic congestion or lockdowns. Consumer goods were sold out because of long and inflexible planning procedures in production and transportation, and empty containers were at the wrong place at the wrong time all over the world. SAP solutions anticipate and mitigate disruptions of this nature:

- SAP Integrated Business Planning (SAP IBP) supports users in understanding the impact of supply chain disruptions and approaches to mitigate them. SAP IBP provides analytics, planning algorithms, what-if simulations, and scenario modeling to stay ahead of any change and improve responsiveness.
- SAP Transportation Resource Planning provides real-time visibility into transportation resources and supports users in finding optimal ways to move empty containers or tank rail cars to ensure availability at minimal cost.
- SAP S/4HANA Supply Chain for transportation management provides functionality for replanning or diverting transportation orders based on estimated or actual time of arrival (ETA and ATA) provided by carriers, and enables seamless integration with transportation service providers.
- SAP Business Network for Logistics is a highly customizable data visualization and reporting platform that comes with intelligent insights functionality. It provides real-time insights into supply chain execution and warnings on supply disruptions from alerts, analytics, and dashboards.

9.9.2 Visibility

On-time delivery at every stage of the transportation chain is important to ensure smooth supply chain processes because buffers shrink as quickly as delays accumulate. Understanding where transportation resources are and when they will be where creates visibility into the transportation process and maximizes the time to react to disruptions in the most efficient way.

- SAP Business Network Freight Collaboration tracks transportation orders based on actual status updates by carriers. It uses a map to visualize the carrier route, unexpected events, and disruptions.

- SAP Business Network, intelligent insights add-on provides situational and performance insights into the global in-transit supply chain. It connects external shipment updates directly to operations plans that depend on the shipment to assess and mitigate the impact of potential or actual delays.

9.9.3 Integration of Trading Partners

More trading partners result in more process complexity. Every small disruption can accumulate and cause bigger problems down the supply chain. Seamless, digital, and real-time integration of these business partners reduces error-prone, paper-based or email communication. It also accelerates and improves processes, giving all partners along the supply chain more visibility and time to respond to exceptions.

- SAP Agricultural Origination Portal enables a seamless and efficient collaboration between farmers and agriculture companies. It provides insights into business transactions such as contracts, deliveries, settlement data, and unsold inventory. This drastically reduces the number of calls and inquiries.
- SAP Business Network Freight Collaboration provides a seamless and real-time integration between shipper companies and their transportation service providers such as freight forwarding companies and carriers. It supports subcontracting, dock appointment scheduling, gate, and settlement processes including evaluated receipt settlement (ERS) and carrier invoicing with exchange of documents and user notes. This solution enables companies to streamline their communication and interaction with their business partners.
- SAP S/4HANA Supply Chain for transportation management enables seamless integration with all trading partners in the supply chain. It offers functionality for generating transportation-relevant documentation and provides regional and mode-specific templates, such as the road waybill for the US and the EU.
- SAP Ariba integrates our customers with their suppliers for processes such as procurement, sourcing, payables, and supply chain.

Chapter 10
The Road Ahead

In this book, we have discussed eight megatrends that are shaping the physical and digital world we live in. Our discussions with customers, partners, and colleagues have touched on additional megatrends that are similarly changing industry boundaries, strategic priorities, business practices, and the human experience.

As we close this book, we want to highlight four *more* megatrends that will continue to change our world. We are looking forward to engaging with our customers and partners in actively shaping our future. (After all, *Business as Unusual with SAP* is about a forward-looking mindset and ongoing conversations, not a cookbook for using SAP solutions to address today's and tomorrow's business priorities.) So we conclude with an outlook on additional megatrends we are observing; we are curious to hear from our readers about your perspectives and points of view:

- **Metaverse**
  Today, digital twins of machines and buildings optimize maintenance and services, but still live their digital lives on digital islands in splendid isolation from the rest of the digital and physical world. On Azeroth, the fantasy setting in the Warcraft franchise, our avatars wield swords and lasers in their quests for gimmicks and glory. Online shopping has become second nature for all of us. Cryptocurrencies seem to be well established, but non-fungible tokens (NFTs) are still puzzling for some or even most of us. Take a quick inventory of the different social media platforms you're subscribed to. We can expect all these digital islands to mesh up, mash up, and form a parallel universe of business, people, and things.
- **Artificial intelligence**
  Evolution hasn't wired our human brains to intuitively understand exponential growth; just think about the old story of covering the 64 fields of a chessboard with rice, starting with one grain on the first square and doubling the number on the next square. Every time we look at artificial intelligence, we find it mostly lackluster. This is because we don't remember how far we have come and because we quickly get used to technology that seemed like science fiction only a few years ago. For example, few people have anticipated that in 2022, machines would paint pictures, compose poems, write novels, drive cars, translate conversations in real time—almost like Douglas Adams' babel fish—or assist with grocery shopping. Perhaps you think the paintings are ugly, the poems weird, the novels bad, the translations error prone, the autonomous cars dangerous, and the conversations stilted,

but if you extrapolate the exponential development of the field of artificial intelligence, you might find the future looks both exciting and a bit scary.

- **Urbanization and future cities**
  The United Nations has estimated that by 2050, almost 70% of the world population of close to 10 billion people will live in urban areas.[1] You don't have to wait for this future to arrive to get an idea of the challenges for mayors and urban planners—just look at the biggest cities today. For example, Tokyo has about the same population as Poland or California. New York City has more citizens than the Netherlands *and* more than the 13 least-populated US states—from New Mexico through Alaska—combined. In those huge agglomerations, how do we keep people safe, employed, connected, educated, entertained, and sustainably housed, mobile, serviced, fed, and watered?
- **Networked business and business networks**
  No business is an island; value chains from raw materials to consumers have always spanned business and industry boundaries. In this book, we have talked about using digitally enabled business networks to support logistics, procurement, vaccine distribution, and data sharing in the automotive industry. But it would be naïve to assume that digitized business processes along value chains only make existing processes more efficient without changing business relationships. Important market mechanisms are based on information asymmetry and speed limits for information transmission. The stock markets show us what happens when differential information access for participants is reduced to microseconds. Digital business networks have a great potential for driving efficiencies, creating and disrupting business models, and changing the dynamics between participants.

Sometimes change happens overnight and disrupts our lives: pandemics break out, financial crises send shockwaves through the global economy, terrorist attacks and wars threaten the world order. Technology transforms business and our lifestyle less abruptly, but sometimes even more profoundly. We have to make our society and economy more resilient against disruption, and we have to ride the technology waves to continuously evolve our business models, shape our business practices, and engage with our fellow humans.

Appendices

Appendix A
Acknowledgments

The 300 pages of this book reflect the voices and perspectives of experts and leaders who have generously shared their experiences with us. We want to cordially thank them for their contributions to *Business as Unusual with SAP: How Leaders Navigate Industry Megatrends*.

Many customers, partners, and market influencers are quoted on these pages and listed in the index. We are grateful for their time and their participation in this unique project:

- Muqsit Ashraf, Accenture
- Tim Baines, Advanced Services Group
- Lars Bolanca, Deutsche Börse Group
- Andy Brown, Galp Energia
- Mark Burton, Bain & Company
- Alejandro Chan, Tetra Pak
- Manuel Cranz, Fressnapf Tiernahrungs GmbH
- Karsten Crede, ERGO Mobility Solutions
- Jeff Johnson, Beckhoff
- Daniel Laverick, Zuellig Pharma
- David Lowson, Capgemini
- Carlos Maurer, Shell
- Thomas Meinel, Evonik
- Aldo Noseda, Eastman
- David Rabley, Accenture
- Coppelia Rose, DXC Technology
- Maia Surmava, Discovery Health
- Daymon Thompson, Beckhoff
- Isabella von Aspern, Stadtwerke Augsburg
- Roland Vorderwülbecke, Gebr. Heinemann
- Anthony Watson, Bank of London
- Marcus Willand, MHP

We appreciate the guidance of many SAP executives who shared their expertise and networks with us:

- Hari Ashvini
- Martin Barkman
- Markus Bechmann
- Senta Belay
- Benjamin Beberness
- Ritu Bhargava
- Michael Byczkowski
- Johnny Clemmons
- Jörg Ferchow
- Rakesh Gandhi
- Stuart Grant
- Hagen Heubach
- Jeff Howell
- Richard Howells

- Stephen Jamieson
- Ralph Kern
- Stephan Klein
- Peter Koop
- Steffen Krautwasser
- Matt Laukaitis
- Mateu Munar
- Adrian Nash
- Mandar Paralkar
- Matt Reymann
- Falk Rieker
- Achim Schneider
- Kevin Schock
- Daniela Sellmann
- Anja Strothkämper
- James Sullivan
- Anton Tomic
- Torsten Welte

Additionally, we want to acknowledge colleagues from SAP who contributed ideas, writing, editing, and encouragement to our project. The book is so much richer because of your involvement, and you have our thanks:

- Kash Al-Aziz
- Chris Atkins
- Peter Beutelmann
- Miquel Carbo
- Eliza Dillard
- Jan Gilg
- Alexander Götz
- Tilman Göttke
- Uwe Grigoleit
- Torsten Hoffmann
- Andrea Kaufmann
- Ulrike Kleifeld
- Martin Klein
- Patrick Lamm
- Sam Masri
- Oliver Nürnberg
- Chris Peck
- Jill Popelka
- Léonie Relic
- Gunther Rothermel
- Frank Ruland
- Jennifer Scholze
- Lowis Seelinger
- Markus Wehrling
- Max Wessel
- Svend Wittern
- Samantha Yerks

We would also like to thank Mark Baven, Vinnie Mirchandani, and Tina Pham of Deal Architect, who spent incalculable time working with the interview content and text in its various iterations.

The final text we offer our readers is a tiny fraction of the hundreds of hours of video conversations and thousands of pages in interview transcripts, presentation slides, and market research reports. We have listed the research we mined in the endnotes.

Appendix B
The Authors

Thomas Saueressig is a member of the Executive Board of SAP SE. He leads the board area SAP Product Engineering and has global responsibility for all business software applications. This includes all functional areas, from product strategy and management to product development and innovation to cloud operations and support. Before this role, Thomas was chief information officer (CIO) and global head of IT services at SAP. He started his career at SAP in consulting for CRM implementations.

Thomas has a degree in business information technology from the University of Cooperative Education in Mannheim (Germany) and a joint executive MBA from ESSEC Business School (France) and Mannheim Business School (Germany).

Peter Maier is president of industries and customer advisory at SAP. In his role, Peter is responsible for making SAP and partner products relevant for SAP customers in their various industries. The customer advisory group showcases SAP solutions and consults SAP customers on achieving the business value generated by the SAP portfolio.

Peter is passionately driving innovation that makes his customers more successful, and he frequently shares his strategy and insights with SAP customers, partners, press, and analysts. He is regularly invited to deliver keynote speeches and to moderate panel discussions. In strategy meetings, he advises leading customers on becoming intelligent and sustainable enterprises in the experience economy.

Peter joined SAP in 1991 as a consultant for the chemicals industry, and has held numerous roles in SAP's industry organization. He is a valued expert on trends, challenges, and best practices in the process industries and beyond, and he has earned the trust of strategic customers as an executive sponsor.

Appendix C
Endnotes

C.1 The End of Business as Usual

[1] Heidi Ledford and Ewan Callaway, "Pioneers of Revolutionary CRISPR Gene Editing Win Chemistry Nobel," *Nature*, October 7, 2020, *http://s-prs.co/5658-0*.

[2] World Bank, "Current Health Expenditure (% of GDP)," World Health Organization Global Health Expenditure Database, January 30, 2022, *http://s-prs.co/56581*.

[3] "The Take: Digital Twins and the Future of Healthcare," SAP News Center, July 5, 2022, *http://s-prs.co/56582*.

[4] "BioNTech Starts Construction of First mRNA Vaccine Manufacturing Facility in Africa," BioNTech (press release), June 23, 2022, *http://s-prs.co/56583*.

[5] Daniel Schmid, "Accelerating Our Commitment to Net-Zero," SAP News Center, January 11, 2022, *http://s-prs.co/56584*.

[6] United Nations Department of Economic and Social Affairs, Energy Statistics Pocketbook 2021 (New York, 2021), *http://s-prs.co/56585*.

[7] Dame Ellen MacArthur, "The Surprising Thing I Learned Sailing Solo around the World," TED, June 29, 2015, *http://s-prs.co/56586*.

[8] Ben Waggoner, "The Cretaceous Period," University of California Museum of Paleontology, November 26, 1995, *http://s-prs.co/56587*.

C.2 Everything as a Service

[1] From Carl Sagan and Ann Druyan, The Demon-Haunted World: Science as a Candle in the Dark (Ballantine Books, 1996), quoted at *http://s-prs.co/56588*.

[2] W. Roy Schulte and Yefim V. Natis, "'Service Oriented' Architectures, Part 1," Gartner, April 12, 1996, *http://s-prs.co/56589*.

[3] Yefim Natis, "SOA Turns 25," Gartner, April 5, 2021, *http://s-prs.co/565810*.

[4] US Bureau of Labor Statistics, "Employment Projections: Employment by Major Industry Sector," September 8, 2022, *http://s-prs.co/565811*.

[5] Vijay Govindarajan and Jeffrey R. Immelt, "The Only Way Manufacturers Can Survive," MIT Sloan Management Review, March 12, 2019, *http://s-prs.co/565812*.

[6] Todd Spangler, "Apple Services Revenue Climbs 17% to Record $19.8 Billion, as iPhone Growth Slows but Still Hits All-Time High," Variety, April 28, 2022, *http://s-prs.co/565813*.

[7] "Rolls-Royce Celebrates 50th Anniversary of Power-by-the-Hour," Rolls-Royce (press release), October 30, 2012, *http://s-prs.co/565814*.

[8] The Advanced Services Group, "Case Studies," date accessed October 7, 2022, *http://s-prs.co/565815*.

[9] General Electric, "TrueChoice Commercial Services," date accessed October 7, 2022, *http://s-prs.co/565816*.

[10] Airbus, "Skywise Digital Solutions," date accessed October 7, 2022, *http://s-prs.co/565817*.

[11] Thomas Lah, "What Is the Fish Model?" Technology & Services Industry Association (TSIA), April 6, 2021, *http://s-prs.co/565818*.

[12] "The Phoebus Cartel," National Public Radio, March 28, 2019, *http://s-prs.co/565819*.

[13] Lloyd O'Donnell, Torsten Welte, Cindy Waxer, "A Circular Path to Sustainability for Manufacturers," SAP Insights, date accessed October 7, 2022, *http://s-prs.co/565820*.

[14] Joe McKendrick, "Right to Repair? Pushback Grows Around Remote Maintenance," RTInsights.com, May 31, 2022, *http://s-prs.co/565821*.

[15] Vinnie Mirchandani, "Burning Platform: The Right to Repair and Service," Deal Architect, May 25, 2022, *http://s-prs.co/565822*.

C.3 Integrated Mobility

[1] Led Zeppelin, 1975, "Kashmir," Track 6 on Physical Graffiti, Swan Song Records, *http://s-prs.co/565823*.

[2] Paul Salopek, "Out of Eden Walk," National Geographic, September 16, 2022, *http://s-prs.co/565824*.

[3] Paul Salopek, "I Was Walking the Global Trail of Our Ancestors. Then Pandemic Struck," National Geographic, May 22, 2020, *http://s-prs.co/565825*.

[4] Ben Jones, "Past, Present and Future: The Evolution of China's Incredible High-Speed Rail Network," CNN Travel, February 9, 2022, *http://s-prs.co/565826*.

[5] Cristina Burack, "VW Plans to Work with China's Didi Ride Service to Design New Car Models," Deutsche Welle (DW), January 5, 2018, *http://s-prs.co/565827*.

[6] "CASE 2.0: A New Framework for The Future of Vehicles and Built Environment," Plug and Play, March 15, 2021, *http://s-prs.co/565828*.

[7] "Connected Car: Its History, Stages and Terms," BMW, October 24, 2019, *http://s-prs.co/565829*.

[8] Zacks Equity Research, "Ford (f) Eliminates End-of-Lease Buyout Options for its EVs," Nasdaq, June 30, 2022, *http://s-prs.co/565830*.

[9] Global Electric Vehicle Sales Up 109% in 2021, with Half in Mainland China," Canalys Newsroom, February 14, 2022, *http://s-prs.co/565831*.

[10] Tara Patel, "Stellantis Warns of Car Market Collapse If EVs Don't Get Cheaper," Bloomberg, June 29, 2022, *http://s-prs.co/565832*.

[11] David Faber, "Exxon Mobil CEO Talks Climate Change with David Faber: Full Interview," CNBC, June 25, 2022 *http://s-prs.co/565833*.

[12] European Federation for Transport and Environment AISBL, Recharge EU: How Many Charge Points Will Europe and Its Members States Need in 2020s, January 2020. *http://s-prs.co/565834*.

[13] European Federation for Transport and Environment AISBL, "Transport and Environment," *http://s-prs.co/565835*.

[14] "Germany's First Mobility 'Flat Rate' Starts in Augsburg," Intelligent Transport, November 4, 2019, *http://s-prs.co/565836*.

[15] BVG Jelbi, "Beliner Verkehrs-Betirbe," last updated February 25, 2021, *http://s-prs.co/565837*.

[16] Catena-X Automotive Network e.V. website, last updated March 2022, *http://s-prs.co/565838*.

[17] Anderson Cooper, "eVTOL: The Flying Vehicles That May Be the Future of Transportation," CBS News, April 17, 2022, *http://s-prs.co/565839*.

C.4 New Customer Pathways

[1] Nadia Reckmann, "What Is Corporate Social Responsibility?" Business News Daily, June 29, 2022, *http://s-prs.co/565840*.

[2] Tom Foster, "Over 400 Startups Are Trying to Become the Next Warby Parker," Inc., May 2018, *http://s-prs.co/565841*.

[3] Casey's General Stores, Inc. History," Funding Universe, date accessed October 20, 2022, *http://s-prs.co/565842*.

[4] Patt Johnson, "Casey's Is Fifth Largest Pizza Chain in U.S.," Des Moines Register, February 16, 2012, *http://s-prs.co/565843*.

[5] Jia Wertz, "Online Sales and the Rise of Returns," Forbes, January 28, 2022, *http://s-prs.co/565844*.

[6] Amazon, "Free Returns," *http://s-prs.co/565845*.

[7] Vinnie Mirchandani, "goTRG's Dominance of Reverse Retail Logistics," New Florence. New Renaissance, September 24, 2021, *http://s-prs.co/565846*.

[8] Vinnie Mirchandani, "Analyst Cam: Specright," Deal Architect, March 9, 2022, *http://s-prs.co/565847*.

[9] Tiffany Burns, Tyler Harris, and Alexandra Kuzmanovic, "The Five Zeros Reshaping Stores," McKinsey & Company, March 16, 2022, *http://s-prs.co/565848*.

[10] Livello GmbH, "Livello Opens Fully Automated 'Smart Store' in Germany Offering 24/7 Unattended Service," Vending Market Watch, January 12, 2021, *http://s-prs.co/565849*.

[11] "We Are Mod," Mod Pizza, *http://s-prs.co/565850*.

[12] "Transform Your Candidate Experience with AI," Paradox.ai, 2019, *http://s-prs.co/565851*.

[13] Yulia Dolzhenkova, "Torsten Toeller: 30th Fressnapf's Anniversary Crowned with Corona," June 17, 2020, ZooInform, *http://s-prs.co/565852*.

[14] "Smartbowl," The One Club, 2016, *http://s-prs.co/565853*.

[15] "For the Love of Dogs, Cats, and the Cloud – How Fressnapf Group Is Shaping an Animal-Centric Ecosystem," Microsoft Customer Stories, May 19, 2021, *http://s-prs.co/565854*.

[16] Vinnie Mirchandani, Silicon Collar: An Optimistic Perspective on Humans, Machines, and Jobs (Deal Architect Inc, September 7, 2016).

[17] "US2612994A Classifying Apparatus and Method," Espacenet Patent Search, *http://s-prs.co/565855*.

C.5 Lifelong Health

[1] Johns Hopkins University Coronavirus Resource Center, "COVID-19 Dashboard," *http://s-prs.co/565856*.

[2] McKinsey & Company, "McKinsey on Healthcare: Perspectives on the Pandemic," *http://s-prs.co/565857*.

[3] "Corona-Warn-App," GitHub, *http://s-prs.co/565858*.

[4] Johns Hopkins Medicine, "Study Suggests Medical Errors Now Third Leading Cause of Death in the U.S.," May 3, 2016, *http://s-prs.co/565859*.

[5] Michael Byczkowski and Magdalena Görtz, "The Industrialization of Intelligence," in Rainer M. Holm-Hadulla, Joachim Funke, and Michael Wink, eds., Intelligence – Theories and Applications, Springer Cham, 2022, *http://s-prs.co/565860*.

[6] McKinsey & Company, "Transforming Healthcare with AU: The Impact on the Workforce and Organizations," March 10, 2020, *http://s-prs.co/565861*.

[7] Steve Lohr, "What Ever Happened to IBM's Watson?" New York Times, July 16, 2021, *http://s-prs.co/565862*.

[8] Michael Byczkowski, "A Promising Prognosis for Healthcare Organizations About Embracing AI," Diagnostics World, February 3, 2022, *http://s-prs.co/565863*.

[9] Bill Siwicki, "Telehealth 2050: The Future Design of Virtual Care Technology," Healthcare IT News, May 10, 2021, *http://s-prs.co/5658170*

[10] National Human Genome Research Institute, "The Human Genome Project," *http://s-prs.co/565864*.

[11] National Human Genome Research Institute, "Personalized Medicine," October 20, 2022, *http://s-prs.co/565865*.

[12] Adrian Gore, "How Discovery Keeps Innovating," McKinsey Quarterly, May 1, 2015, *http://s-prs.co/565866*.

[13] Adrian Gore, "How Discovery Keeps Innovating," McKinsey Quarterly, May 1, 2015, *http://s-prs.co/565866*.

[14] ECA Foundation (Mannheim), "EU GMP Annex 16: Certification by a Qualified Person and Batch Release," ECA Academy, *http://s-prs.co/565867*.

[15] "BioNTech Introduces First Modular mRNA Manufacturing Facility to Promote Scalable Vaccine Production in Africa," BioNTech (press release), February 16, 2022, *http://s-prs.co/565868*.

[16] Magdalena Görtz, Michael Byczkowski, and Mathias Rath "A Platform and Multisided Market for Translational, Software-Defined Medical Procedures in the Operating Room (OP 4.1): Proof-of-Concept Study," JMIR Medical Informatics, vol. 10, no. 1 (January 2022), *http://s-prs.co/565869*.

[17] The World Bank, "Life Expectancy at Birth, Total (Years)," *http://s-prs.co/565870*.

[18] Centers for Medicare & Medicaid Services, "National Health Expenditure Data," *http://s-prs.co/565871*.

[19] World Health Organization, "The Top 10 Causes of Death," December 9, 2020, *http://s-prs.co/565872*.

C.6 The Future of Capital and Risk

[1] The Big Short, directed by Adam McKay (Paramount Pictures, 2015).

[2] Hillary Jackson, "Investopedia's Oddest Business and Investing Terms," Investopedia, December 22, 2021, *http://s-prs.co/565874*.

[3] Report Linker, "Financial Services Global Market Report 2022," May 2022, *http://s-prs.co/565875*.

[4] "Gartner Says Digitalization Will Make Most Heritage Financial Firms Irrelevant by 2030," Gartner, October 29, 2018, *http://s-prs.co/565876.*

[5] "Resolvability Assessment of Major UK banks: 2022," Bank of England, June 10, 2022, *http://s-prs.co/565877.*

[6] Stephen Deane, "The Three Big Transitions Reshaping Finance," Barron's, September 2, 2021, *http://s-prs.co/565878.*

[7] Goldman Sachs, "Marcus by Goldman Sachs Leverages Technology and Legacy of Financial Expertise in Dynamic Consumer Finance Platform," date accessed October 10, 2022, *http://s-prs.co/565879.*

[8] Falk Rieker, "How Behavioral Banking Can Drive Financial Literacy and Inclusion," Forbes, September 3, 2020, *http://s-prs.co/565880.*

[9] "Crypto Banks," Banks.com, *http://s-prs.co/565881.*

[10] Oswego State University of New York, "The Basics about Cryptocurrency," date accessed October 10, 2022, *http://s-prs.co/565882.*

[11] "Rabobank Sustainable Funding Framework," Rabobank, June 2021, *http://s-prs.co/565883.*

[12] Judith Magyar, "How Vast Bank Makes It Easy to Invest in Cryptocurrency," Forbes, September 23, 2021, *http://s-prs.co/565884.*

[13] "The Bank of London Launches Today as the World's First Purpose-built Global Clearing, Agency & Transaction Bank," Cision PR Newswire, November 30, 2021, *http://s-prs.co/565885.*

[14] James Chen, "Unicorn: What It Means in Investing, With Examples," Investopedia, May 31, 2022, *http://s-prs.co/565886.*

[15] Simon Taylor, "Bonus Food: Ping-An, the tech giant masquerading as a finance company," Fintech Brain Food, October 8, 2020, *http://s-prs.co/565887.*

[16] W. Chan Kim, Renèe Mauborgne, and Mi Ji, "Ping An Good Doctor: Creating a Non-disruptive Solution for China's Healthcare System," Insead Publishing, February 25, 2021, *http://s-prs.co/565888.*

[17] Caroline Banton, "Reinsurance," Investopedia, April 3, 2022, *http://s-prs.co/565889.*

[18] "SAP S/4HANA Insurance for reinsurance management," SAP Help Portal, date accessed October 10, 2022, *http://s-prs.co/565890.*

[19] Karsten Crede, "Boxing Mid-Range – 3 Years of ERGO Mobility Solutions," LinkedIn, August 3, 2020, *http://s-prs.co/565891.*

[20] "ERGO Mobility Solutions as Insurance Orchestrator in Mobility Ecosystem," DYCSI, July 29, 2022, *http://s-prs.co/565892.*

[21] ERGO, "Mobility Technology Center in Munich established," May 19, 2022, *http://s-prs.co/565893.*

C.7 Sustainable Energy

[1] National Aeronautics and Space Administration (NASA), "Graphic: The Relentless Rise of Carbon Dioxide," *http://s-prs.co/565894*.

[2] Edward Franklin Degering, Carl Bordenca, Bernard Genry Gwynn, An Outline of Organic Nitrogen Compounds (Swift, 1942).

[3] Kate Abnett, "EU Parliament Backs Labelling Gas and Nuclear Investments as Green," Reuters, July 6, 2022, *http://s-prs.co/565895*.

[4] David Faber, "Exxon Mobil CEO Talks Climate Change with David Faber: Full Interview," CNBC, June 25, 2022, *http://s-prs.co/565896*.

[5] Vivek Chidambaram, et al., "The Changing Joule Dynamic: Five Portfolio Plays to Help Oil and Gas Companies Win in the New Energy Transition," Accenture, last accessed October 21, 2022, *http://s-prs.co/565897*.

[6] "Global Direct Primary Energy Consumption," Our World in Data, last accessed October 20, 2022, *http://s-prs.co/565898*.

[7] "DNV 2022 Energy Transition Outlook: A Global and Regional Forecast to 2050," Det Norske Veritas, *http://s-prs.co/5658172*.

[8] Technik Museum Sinsheim, Growian Wind Turbine Blade, *http://s-prs.co/565899*.

[9] Matthew Dalton, "Behind the Rise of U.S. Solar Power, a Mountain of Chinese Coal," The Wall Street Journal, July 31, 2021, *http://s-prs.co/5658100*.

[10] Hannah Ritchie, Max Roser, and Pablo Rosado (2020), "Energy." *http://s-prs.co/5658101*.

[11] DeStatis Statistisches Bundesamt, "Electricity Generation in the First Half of 2022: 17.2% More Electricity from Coal than in the Same Period of the Previous Year," September 7, 2022, *http://s-prs.co/5658102*.

[12] Umwelt Bundesamt, "Primary Energy Consumption," *http://s-prs.co/5658103*.

[13] Forbes Breaking News, "Which Uses More Electricity...A Refrigerator When It's Running or Electric Car When It's Charging?" YouTube, July 19, 2022, *http://s-prs.co/5658104*.

[14] Michael Wayland, "Stellantis CEO Warns of Electric Vehicle Battery Shortage, Followed by Lack of Raw Materials," CNBC.com, May 24, 2022, *http://s-prs.co/5658105*.

[15] Simon P. Michaux, Assessment of the Extra Capacity Required of Alternative Energy Electrical Power Systems to Completely Replace Fossil Fuels, GTK Open File Work Report 42/2021, Geological Survey of Finland, 2021, *http://s-prs.co/5658106*.

[16] Vinnie Mirchandani, "Supercritical Geothermal Energy," New Florence. New Renaissance, April 8, 2022, *http://s-prs.co/5658107*.

[17] Samuel K. Moore, "Gravity Energy Storage Will Show Its Potential in 2021," IEEE Spectrum, January 4, 2021, *http://s-prs.co/5658108*.

[18] World Economic Forum, The Net-Zero Industry Tracker, July 28, 2022, *http://s-prs.co/5658109*.

[19] "Greenhouse Gas Protocol," Greenhouse Gas Protocol, date accessed October 21, 2022, *http://s-prs.co/5658110*.

[20] The International Council on Clean Transportation, Carbon Intensity of Crude Oil in Europe: Executive Summary, 2010, *http://s-prs.co/5658111*.

[21] Ian Tiseo, "Passenger Car Carbon Dioxide Emissions Worldwide 2010-2020," Statista, December 14, 2021, *http://s-prs.co/5658112*, and Ian Tiseo, "Global CO_2 Emissions 1970-2020, by Sector," Statista, September 9, 2022, *http://s-prs.co/5658113*.

[22] Office of Energy Efficiency and Renewable Energy, "Pumped Storage Hydropower," US Department of Energy, *http://s-prs.co/5658114*.

[23] Luis "Nando" Ochoa, et al, "Project EDGE," The University of Melbourne Department of Electrical and Electronic Engineering, August 9, 2022, *http://s-prs.co/5658115*.

[24] Nick Geary, "Somewhere over the Hydrogen Rainbow," Boiler Guide, July 6, 2022, *http://s-prs.co/5658116*.

[25] David Kindy, "Fossil Fuel—Free 'Green' Steel Produced for the First Time," Smithsonian Magazine, August 31, 2021, *http://s-prs.co/5658117*.

[26] Shell, "Major Innovation Milestones," *http://s-prs.co/5658119*.

[27] Shell, "Powering Progress," *http://s-prs.co/5658120*.

[28] Shell, "The Energy Transformation Scenarios," *http://s-prs.co/5658121*.

[29] Shell, "Shell Scenarios," *http://s-prs.co/5658122*.

[30] World Economic Forum, "Clean Skies for Tomorrow Coalition," *http://s-prs.co/5658173*.

[31] Shell, "Energy Transition Progress Report 2021," *http://s-prs.co/5658125*.

[32] Johan Krebbers, "A Transparent Carbon Footprint of Supply Chains," Shell, July 7, 2021, *http://s-prs.co/5658126*.

[33] Shell, "Map Your Decarbonisation Journey with Us," *http://s-prs.co/5658127*.

[34] Peter Maier, "Decarbonizing Industry Value Chains Is a Team Effort," SAP News Center, July 8, 2022, *http://s-prs.co/5658128*.

[35] European Commission, "Renewable Energy Targets," European Union, *http://s-prs.co/5658129*.

C.8 Circular Economy

[1] David Latchman, "Deciphering the Science and Linguistics of 'Arrival,'" Science vs. Hollywood, December 13, 2016, *http://s-prs.co/5658130*.

[2] Thoth Adan, "Symbols Based on Circles," January 15, 2019, *http://s-prs.co/5658131*.

[3] Shawn Martin, "The Age of Erasable Books," The Atlantic, July 9, 2014, *http://s-prs.co/5658132*.

[4] Hinton's Waste, "History of Recycling [Timeline]," *http://s-prs.co/5658133*.

[5] The Aluminum Association "Recycling," *http://s-prs.co/5658134*.

[6] "A Fountain of Youth," Porsche Newsroom, June 26, 2014, *http://s-prs.co/5658135*.

[7] "Recycling Rate of Metals and Glass Worldwide as of 2018, by Region," Statista, January 13, 2022, *http://s-prs.co/5658136*.

[8] H. Hauggaard-Nielsen, M. Gooding, P. Ambus, et al., "Pea–Barley Intercropping for Efficient Symbiotic N2-Fixation, Soil N Acquisition and Use of Other Nutrients in European Organic Cropping Systems, Field Crops Research, vol. 113, no. 1 (July 2009), *http://s-prs.co/5658137*.

[9] Scott Russell, "Reimagining the Fashion Industry by Designing Out Waste," SAP News Center, October 25, 2021, *http://s-prs.co/5658138*.

[10] Alyssa Danigelis, "Procter & Gamble Patent Hints at Recyclable Diapers, Absorbent Products," Environment + Energy Leader, July 3, 2019, *http://s-prs.co/5658139*.

[11] Erin Blakemore, "The Shocking River Fire That Fueled the Creation of the EPA," History.com, December 1, 2020, *http://s-prs.co/5658140*.

[12] Tensie Whelan and Randi Kronthal-Sacco, "Research: Actually, Consumers Do Buy Sustainable Products," Harvard Business Review, June 19, 2019, *http://s-prs.co/5658141*.

[13] Merrit Kennedy, "Lead-Laced Water in Flint: A Step-By-Step Look at the Makings of a Crisis," NPR, April 20, 2016, *http://s-prs.co/5658142*.

[14] Carbon Trust, "Briefing: What Are Scope 3 Emissions?" *http://s-prs.co/5658143*.

[15] "SAP Unilever Pilot Blockchain Technology Supporting Deforestation-Free Palm Oil," Unilever (press release), March 20, 2022, *http://s-prs.co/5658144*.

[16] Vivek Bapat, "Allbirds' Runaway Success Leaves a Small Carbon Footprint," SAP News Center, July 2, 2021, *http://s-prs.co/5658145*.

[17] "Eastman Enters Exclusive Negotiation with Site in Normandy for Molecular Recycling Facility in France," Eastman (press release), March 29, 2022, *http://s-prs.co/5658146*.

[18] "Mechanical and Molecular Recycling—Some Things Are Just Better Together," Eastman, *http://s-prs.co/5658147*.

[19] "Welcome to the Renew World," Eastman, *http://s-prs.co/5658148*.

[20] Vinnie Mirchandani, "Analyst Cam: Algramo – Redefining CPG Plastic Packaging," Deal Architect, March 2, 2022, *http://s-prs.co/5658149*.

[21] Vinnie Mirchandani, "Analyst Cam: Limeloop – Circular Economy Logistics," Deal Architect, February 23, 2022, *http://s-prs.co/5658150*.

[22] Ellen MacArthur Foundation, "About Us," *http://s-prs.co/5658151*.

[23] Ellen MacArthur Foundation, "The Butterfly Diagram: Visualising the Circular Economy," *http://s-prs.co/5658152*.

[24] Ellen MacArthur Foundation, "Vision of a Circular Economy for Fashion," *http://s-prs.co/5658153*.

[25] Ellen MacArthur Foundation, "Developing Regenerative Chemistry: Solvay," YouTube, *http://s-prs.co/5658154*.

C.9 Resilient Supply Networks

[1] Vincent A. Mabert and Michael J. Showalter, "Logistics of the American Circus: The Golden Age," Production and Inventory Management Journal, 2010, *http://s-prs.co/5658155*.

[2] Richard Howells, "Baby Formula Shortage Underscores Need for Supply Chain Resilience," Forbes, Mary 19, 2022, *http://s-prs.co/5658156*.

[3] Richard Howells, "When Semiconductor Chips Are Down, What Can Be Done?" Forbes, January 26, 2022, *http://s-prs.co/5658157*.

[4] Richard Howells, "Container Ship Crisis Underscores Critical Supply Chain Crisis," Forbes, October 7, 2021, *http://s-prs.co/5658158*.

[5] Vinnie Mirchandani, "Amazon's Largest Fulfillment Center," New Florence. New Renaissance, November 29, 2021, *http://s-prs.co/5658159*.

[6] Chad P. Bown and Thomas J. Bollyky, "How COVID-19 Vaccine Supply Chains Emerged in the Midst of a Pandemic," Peterson Institute for International Economics, August 2021, *http://s-prs.co/5658160*.

[7] "Pfizer-BioNTech COVID-19 Vaccine: US Distribution Sheet," *http://s-prs.co/5658161*.

[8] National Retail Federation, "Customer Returns in the Retail Industry," January 25, 2022, *http://s-prs.co/5658162*.

[9] Vinnie Mirchandani, "goTRG's Dominance of Reverse Retail Logistics," New Florence. New Renaissance, September 24, 2021, *http://s-prs.co/5658163*.

[10] Stefan Thomke, "Mumbai's Models of Service Excellence," Harvard Business Review, November 2012, *http://s-prs.co/5658164*.

[11] "Size of the Online Food Delivery Market across India from 2016 to 2020," Statista, *http://s-prs.co/5658165*.

[12] SAPVideoMOM, "Hannover Messe 2022 - Experience Design to Operate for Discrete Manufacturers," YouTube, June 2, 2022, *http://s-prs.co/5658166*.

[13] SAPVideoMOM, "Hannover Messe 2022 - Experience Design to Consume for Process Manufacturers," YouTube, June 2, 2022, *http://s-prs.co/5658167*.

[14] SAP, "Maintaining Business Continuity During Periods of Semiconductor Shortages," SAP Thought Leadership Paper, 2022, *http://s-prs.co/5658168*.

C.10 The Road Ahead

[1] United Nations, "68% of the World Population Projected to Live in Urban Areas by 2050, Says UN," May 16, 2018, *http://s-prs.co/5658169*.

Index

A

B

C

D

E

F

G

H

I

J

K

L

M

T

U

V